Frontier Road

T0201146

Antipode Book Series

Series Editors: Vinay Gidwani, University of Minnesota, USA and Sharad Chari, CISA at the University of the Witwatersrand, USA

Like its parent journal, the Antipode Book Series reflects distinctive new developments in radical geography. It publishes books in a variety of formats – from reference books to works of broad explication to titles that develop and extend the scholarly research base – but the commitment is always the same: to contribute to the praxis of a new and more just society.

Published

Frontier Road

*Power, History, and the Everyday State
in the Colombian Amazon*

Simón Uribe

WILEY Blackwell

Registered Offices
John Wiley & Sons, Inc., 111 River Street, Hoboken, NJ 07030, USA
John Wiley & Sons Ltd, The Atrium, Southern Gate, Chichester, West Sussex, PO19 8SQ, UK

Editorial Office
9600 Garsington Road, Oxford, OX4 2DQ, UK

For details of our global editorial offices, customer services, and more information about Wiley products visit us at www.wiley.com.

Wiley also publishes its books in a variety of electronic formats and by print-on-demand. Some content that appears in standard print versions of this book may not be available in other formats.

Library of Congress Cataloging-in-Publication Data

Names: Uribe, Simón, author.
Title: Frontier road : power, history, and the everyday state in the Colombian Amazon / Simón Uribe.
Description: Hoboken, NJ : John Wiley & Sons, 2017. | Includes bibliographical references and index.
Identifiers: LCCN 2016044212| ISBN 9781119100171 (cloth) | ISBN 9781119100188 (pbk.)
Subjects: LCSH: Roads–Colombia–Putumayo (Department) | Infrastructure (Economics)–Colombia–Putumayo (Department) | Roads–Design and construction.
Classification: LCC H359.C7 U75 2017 | DDC 338.9861/63–dc23
LC record available at https://lccn.loc.gov/2016044212

Cover Images: Image 1: Lorry making the route between Mocoa and San Francisco, c. 1950 (Reproduced by permission of the Archive of the Diocese of Sibundoy)
Image 2: San Francisco-Mocoa road © Simón Uribe, 2010
Cover Design: Wiley

Set in 10.5/12.5pt Sabon by SPi Global, Pondicherry, India
Printed and bound in Malaysia by Vivar Printing Sdn Bhd

10 9 8 7 6 5 4 3 2 1

To Antonio, and to the memory of Roberto Franco
and Guillermo Guerrero

Contents

Series Editors' Preface

The *Antipode Book Series* explores radical geography 'antipodally,' in opposition, from various margins, limits or borderlands.

Antipode books provide insight 'from elsewhere,' across boundaries rarely transgressed, with internationalist ambition and located insight; they diagnose grounded critique emerging from particular contradictory social relations in order to sharpen the stakes and broaden public awareness. An *Antipode* book might revise scholarly debates by pushing at disciplinary boundaries, or by showing what happens to a problem as it moves or changes. It might investigate entanglements of power and struggle in particular sites, but with lessons that travel with surprising echoes elsewhere.

Antipode books will be theoretically bold and empirically rich, written in lively, accessible prose that does not sacrifice clarity at the altar of sophistication. We seek books from within and beyond the discipline of geography that deploy geographical critique in order to understand and transform our fractured world.

Vinay Gidwani
University of Minnesota, USA

Sharad Chari
CISA at the University of the Witwatersrand, USA

Antipode Book Series Editors

Acknowledgements

Several people and institutions have supported me through the long process of completing this book. Fieldwork and archive work were conducted in Barcelona, Bogotá and Putumayo from 2009 to 2011, and was funded with research grants from the Wenner-Gren Foundation, the London School of Economics, the University of London and the Abbey-Santander Travel Research Fund. During this period, many people contributed directly or indirectly to the research. I would like to express my deep gratitude and indebtedness to all of them, including those whom I may forget to mention here.

In the Putumayo, I owe special thanks to Judy and Guillermo Guerrero, Don Hernando Córdoba and his family, Doña Ruth, Humberto Toro, Franco Romo, Gerardo Rosero, Narciso Jacanamejoy, María Cerón, Humberto Tovar, Elvano Camacho, Rigoberto Chito, Guillermo Martínez, Mauricio Valencia, Guido Revelo, Silvana Castro, Felipe Arteaga, Adriana Barriga, Jorge Luis Guzmán, Bernardo Pérez and Gladys Bernal, Edgar Torres, and Alejandro and Rocío Ortiz.

In Barcelona, I want to thank Fra Valentí Serra, who granted me access to the Provincial Archive of the Capuchins of Catalonia (APCC), a rich source for the history of the road; and also to Lina González and Santiago Colmenares for their great hospitality and comradeship. The archive work in Barcelona was complemented by research in the Archive of the Diocese of Sibundoy in Putumayo (ADS), possible thanks to the help of Gustavo Torres; and in the National Library and the National Archive in Bogotá (AGN), carried out with the assistance of María Elisa Balen and Joaquín Uribe. In New York, where I spent an academic semester as an exchange student in the Department of Anthropology at Columbia University, I was fortunate to have the guidance of Michael Taussig, who offered generous advice and also introduced me to Timothy Mitchell and Richard Kernaghan, both of whom gave me useful insights during the early stages of the project. I would also like to thank Bret Ericson, Nando, Nicolás Cárdenas and Orlando Trujillo, who made my stay in New York enjoyable.

The bulk of the writing was done between 2011 and 2013, and was funded with a writing grant from the Foundation for Regional and Urban Studies (Oxford) and a scholarship from Colciencias (Bogotá). During this time, I received academic advice and personal support from several people. In the UK, I am especially indebted to Sharad Chari and Gareth Jones, who provided continuous guidance and support throughout my PhD research, which forms the basis of much of the book. In Colombia, Stefania Gallini and the Environmental History research group, Augusto Gómez, María Clemencia Ramírez, Martha Herrera and the members of the Umbra research workshop, offered valuable feedback during the writing process. Last but not least, posthumous thanks and appreciation go to my friend Roberto Franco, who first awoke my interest in the Amazon region and its history.

The people at Wiley-Blackwell did a brilliant job in turning a raw manuscript into a finished book. Two anonymous reviewers meticulously read the different versions of the manuscript, providing thoughtful comments and critiques. Jacqueline Scott and the series editors provided efficient and generous guidance throughout the process. I want to express my thanks to them, as well as to the different persons who collaborated in the different stages of the edition and production process.

Finally, my deep gratitude goes to my friends and family, who supported and endured me all the way. And, of course, to María Elisa, for her company and unconditional help; in numerous ways this book is hers as well.

Introduction

The 148 kilometres that separate Mocoa from Pasto are terrifying. So say the drivers that daily cross the *páramos*,[1] valleys and inhospitable *selvas* along the road between the two cities, a journey that can take up to 10 or 12 hours and sometimes much longer depending on the state of the road or the action of the guerrilla … This is the road traversed by the conqueror Hernán Pérez de Quesada, who defied the abysses, *páramos* and numerous water courses that criss-cross it, accompanied by 270 soldiers, 200 horses and ten Indians that guided him in the conquest of the south. It was also the route that by 1835 was used by merchants eager to arrive at the Putumayo River to transport rubber, quinine and *tagua* by canoe to Manaus and Belen de Para and to return with iron, salt, liqueurs and other foreign goods.

On account of the obstacles this road imposes on travel to the Putumayo, General Rafael Reyes turned Mocoa into a prison and there exiled his political enemies. This road was also traversed by the Colombian troops who defended the national sovereignty during the conflict with Peru in 1932 … Through this same road came the stream of *colonos* on the pretext of transforming the region; and also those who fled political violence, immigrants attracted by the discovery of oil, and finally those deluded with the *coca* boom.

To get in or out of this region is uncertain … For this reason [drivers] do not hesitate to have a drink of *aguardiente* in order to control their nerves and face the fractured rocks, slopes flowed [sic] with high pressure water, creeks and brooks, and a dense mist that makes this place a world apart.
'Pasto-Mocoa road: 148 km of fear' (*El Tiempo*, 3rd November 1996).

Frontier Road: Power, History, and the Everyday State in the Colombian Amazon,
First Edition. Simón Uribe.
© 2017 John Wiley & Sons Ltd. Published 2017 by John Wiley & Sons Ltd.

This is one of the many depictions of a road connecting the Andean and Amazon regions in southwest Colombia (see Figure I.1), infamously known as *El trampolín de la muerte* [the *trampoline of death*] due to its sheer and precipitous topography. These depictions appear from time to time in the national press, travellers' blogs, YouTube videos and TV news reports. On the occasion that a bus falls off a cliff or is buried under an avalanche, leaving a death toll of more than 10 or 20, or when travellers are trapped in landslides and have to be air-lifted, these descriptions multiply. During such events, condemnations and promises proliferate: journalists portray horrific scenes of mud, wreckage, blood and unfound corpses while reiterating the archaic state of the road; locals lay blame on the government for perpetual neglect; the president announces the imminent launch of a long-awaited road project that will finally redeem a country's rich yet forgotten margin of the state; politicians accuse each other while promising a 'definite solution' if they are elected. Repetition turns each tragedy into farce, as characters re-play the same script, replicating the staple fare of the frontier: isolation, confinement, violence, lawlessness, backwardness, abandonment, neglect, terror and fear.

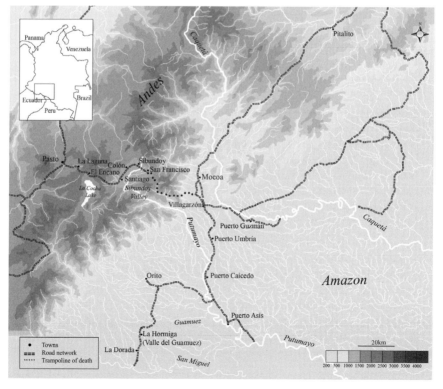

Figure I.1　Colombia's Andean-Amazon region.

Through reiteration and replication, this vocabulary has become indissoluble from the geographical imagery of the road, affixed to the various names by which it has been baptized ('wages of fear', 'the longest cemetery in the world', 'shortcut to hell', 'the dumb death'). The most popular of these terms remains the *trampoline of death*, which sharply captures the sense of being under constant threat of plunging into a bottomless void. Each of these names, together with the written and visual accounts they echo, conveys the striking features of this infrastructural landscape: its almost impossible layout, which from the distance looks like a thin, meandering path carved in a vertical forest; the palpable fragility and instability of the entire infrastructure, denoted by all sorts of 'danger' and 'caution' signs and evinced by persistent landslides wearing away the road surface, crumbling slopes and culverts eroded or collapsed by the action of water; its unsettling atmosphere, composed by the coming and going of roaring engines muffled by thick masses of fog crawling up the *cordillera*; and the ubiquitous remnants of deadly events, differently marked with plaques, shrines and fragments of debris scattered throughout the road.

To traverse the *trampoline of death*'s exceptional landscape would most probably make the traveller feel that he is inhabiting a 'world apart', as the journalist euphemistically puts it. Still, for the inhabitants of regions traditionally deemed as peripheral, isolated, excluded from or yet to be assimilated into the state, regions most commonly known in official and academic language as *fronteras internas* (internal frontiers), infrastructures like the *trampoline of death* have long been the norm rather than the exception. In Colombia, where the sum of these regions is still variously estimated to comprise from three-quarters to one-half of the country's total area, such infrastructures, commonly branded as *trochas* (trails), abound, their ruinous and neglected state often projected to the entire territory and population they encompass. This image is similarly echoed in the frontier, where these infrastructures are heavily invested with enduring feelings and memories of isolation, exclusion and abandonment from the state. Inversely, the building of smooth paved roads annihilating spatial barriers and shrinking geographical distances constitutes an everyday expectation, one that powerfully embodies the long-awaited promise of development, progress and inclusion.

The evocative power that roads have as physical structures that express feelings and visions of modernity, backwardness, abjection or development, has been widely stressed.[2] This affective dimension of roads is especially manifest in 'peripheral' or 'marginal' spaces, where they are conspicuous by their incomplete or precarious state.[3] Precariousness and incompleteness, however, do not undermine the vital role roads have played in the history of these regions. This role is both related to their function as intrinsic technologies of state-building and to their singular significance in such spaces,

where they have been customarily seen as infrastructures aimed at symbolically and physically civilizing 'savage' or 'backward' lands and populations
through the interwoven ends they are meant to assist and achieve: colonization, sovereignty, legibility, security and development.[4]

This view prevailed for many years in scholarly accounts of the frontier
in the Amazon, where they came to be regarded as a primary means to
materialize popular slogans such as 'land without men for men without
land'.[5] The racial, environmental and social violence that this image
sustained has been amply documented and criticized, throwing light on
the conflicts shaping frontier processes throughout the region.[6] The road
from Pasto to Puerto Asís, of which the *trampoline of death* is one of
several fragments (Figure I.1), provides a clear example of the state's
civilizing project and the violence this rhetoric has historically sustained.
This violence can be traced through the road's many characters, conflicts
and events, as well as in the entangled political and social dynamics it has
assisted. Although this violence has not deprived the road of its promise
of connection and inclusion, it has revealed the political economies and
ecologies of infrastructural development region wide. More significantly,
this violence speaks of the spatio-temporal process of state-building and
of the role the frontier has played throughout it.

Frontier Road critically examines this process through an ethnographic
and historical exploration of this singular infrastructure, from its inception
in the nineteenth century to the present and through its various shapes and
transfigurations: indigenous and *cauchero* (rubber tapper) trail, missionary
bridle path, colonization dirt road and interoceanic megaproject. In reconstructing this history, I show how the Colombian Amazon was constituted
and assimilated into the order of the state as a frontier space and, in turn,
how this condition of frontier became vital to the existence of this order.
In this sense, I argue that this territory has never been excluded from the
spatial and political order of the state, but rather incorporated to this
order through a relationship of inclusive exclusion. The meaning and
nature of this relationship, to be discussed later in this chapter, confronts
traditional notions of the state and the frontier. Yet the purpose of the
book, as I hope will become clear in due course, is not just to question
such notions, but also, and more importantly, to expose how they have
helped legitimate a hegemonic political, social and spatial order.

Colombia's amputated map

Among the various connotations of the term 'frontier' (territorial or
national boundary, zone of contact between different cultures, fringe of
settled areas, safety valve), one of the most lasting connotations has been

that of wild and untamed spaces embodying the antithesis of civilization. This image has long pervaded representations of the Colombian Amazon and other 'internal frontiers', consistently portrayed in the media as *no man's lands* or stateless territories occupied and controlled by subversive or outlaw forces.[7] Most significantly, this constitutes an image that constantly surfaces in the historiography when elucidating Colombia's 'unfinished' or 'failed' project of nation building. This is particularly the case of the scholarship concerned with the country's long history of violence and political conflict, the origins and persistence of which tend to be explained in terms of a 'fragmented', 'weak', 'precarious', 'absent' or 'co-opted' state.[8] These adjectives are especially pervasive when alluding to internal frontiers, and regularly overlap with moral ones so that isolation and neglect are conflated with backwardness, lawlessness and violence.

When seen from a long-term perspective, this view is often linked to the broader premise that the country's geography constitutes a key factor explaining the singular features of its economic, social, and political history. Expressions such as 'fragmentation', 'isolation', 'atomization', 'dispersion' and 'complexity' form part of a shared vocabulary used within the historical and geographical literature to depict the manifold direct or indirect, and mostly negative, influences of geography on the country's historical development.[9]

An illustrating example of a geographical approach to Colombia's history can be found in *Colombia. Fragmented land, divided society* (Safford and Palacios 2002), a reference book that provides a condensed historical account of the country from pre-Columbian times to the late-twentieth century. The burden of geography on the country's history, strongly emphasized in the book's title and its cover (showing the gloomy portrait, typical of nineteenth-century iconography, of a White traveller carried through the *cordillera* on the back of a *sillero*, see Figure I.2) is summarized at the beginning of the introduction as follows:

Colombia's history has been shaped by its spatial fragmentation, which has found expression in economic atomization and cultural differentiation. The country's historically most populated areas have been divided by its three mountain ranges, in each of which are embedded many small valleys. The historical dispersion of much of the population in isolated mountain pockets long delayed the development of transportation and the formation of an integrated national market. It also fostered the development of particularized local and regional cultures. Politically, this dispersion has manifested itself in regional antagonism and local rivalries, expressed in the nineteenth century in civil war and in at least part of the twentieth century in intercommunity violence (Safford and Palacios 2002, p. ix).

Figure I.2   'The mount of Agony', engraved by Émile Maillard after a sketch by André and Riou.
Source: André 1877, p.363.

Throughout the book, the authors strongly emphasize the relationship between the country's spatial and political fragmentation, and how this situation has historically been both a cause and a reflection of Colombia's long-standing difficulties in attempting to build a solid nation state. The internal frontiers, on the other hand, are conspicuous by their absence, except when it comes to stressing the violent dynamics associated with them, or with their marginal significance within the country's history. One of the few references made to them in the text, for instance, reads: 'Colombia's other great forested region, the Putumayo and Amazonia, was visited by few Spanish-speaking Colombians until the twentieth century. And even now these regions are only partly integrated to the national polity and economy' (Safford and Palacios 2002, p.9); another alludes to the boom of extractive economies at the dawn of the twentieth century: 'In more than half of the territory of Colombia, a violent frontier society emerged of which the national state had little knowledge and over which it had even less control' (p.278).

Even more telling is the map that accompanies the book's introduction (Figure I.3), where such territories are partly removed or dissected, partly shown blank and otherwise filled with the map conventions. This amputated map, different versions of which can be found reproduced indefinitely in official atlases and history textbooks, strongly reflects and reinforces the dominant image of the frontier as vast peripheral zones falling within the country's geographical borders yet lying beyond the limits of the state.

The prevailing, and seemingly obvious, answer to the question of why a significant portion of the country still constitutes an internal frontier, is that the state has historically been too weak or simply unwilling to reach and control its peripheral regions. As noted, this constitutes an explanation in which geography is given a great causal weight, and which manifests itself in statements such as 'Colombia tiene más geografía que estado' [Colombia's geography surpasses the state] (cited in García and Espinosa 2011, p.53), an often quoted expression from the former vice president Gustavo Bell. This explanation, moreover, largely stems from a tendency to conceive the state in a Weberian way or – as the classical definition goes – as 'a human community that (successfully) claims the *monopoly of the legitimate use of physical force* within a given territory' (Weber 1998, p.78, emphasis in original). Put differently, in this view the state's degree of 'success' – or failure – is measured against its capacity to exert *physical* control or domination over a given territory.

In accordance with this view, the relation between the frontier and the state is perceived as a sort of zero-sum game where the expansion of one is expressed in the contraction of the other or – in Ratzelian terms (1896) – as an 'organic' outward movement from centre to periphery against which the strength of the state is measured. The preservation and proliferation of all sorts of frontiers in the body politic of nation-states, suggests however that the former constitute spaces whose role is central for the very existence of the state (Serje 2011; Hansen and Stepputat 2005; Das and Poole 2004). In what follows, I relate this role with the notion of exception as a way to elucidate its nature and show how it leads to a different understanding of the state.

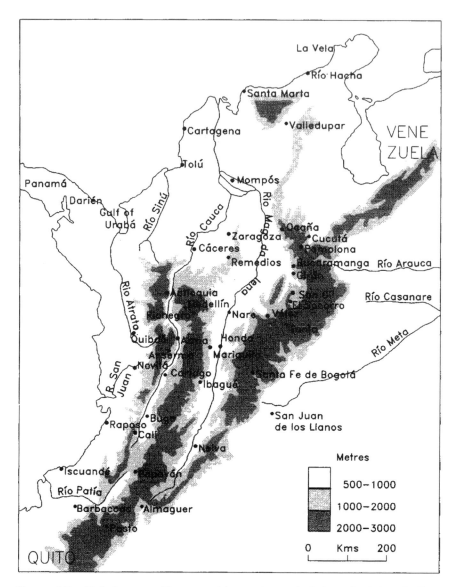

Figure I.3   'Relief map, with some cities at the end of the colonial period'.
Source: Safford and Palacios 2002, p.2, quoted from McFarlane 1993, p.11.

The frontier as a space of exception

If we see the frontier as a space that does not lie outside the order of the
state and yet at the same time a space that is by definition opposed and
external to this order or, at a broader level, as a condition of *'being
outside, and yet belonging'* (Agamben 2005, p.35), the question is how

to situate this space within the architecture of this order. In addressing this question, we need to look at the relationship between exception, sovereign power and violence. This relationship was first theorized by Carl Schmitt (1985 [1922]), who argued that the legal figure of the state of exception is a crucial mechanism to guarantee the existence of the state. The main premise underlying this argument, which underpins the German legal-theorist critique of liberal constitutionalism, is that the integrity of the state is constantly threatened by situations of conflict and disorder. As such situations cannot be totally anticipated and hence legally prescribed, the sovereign, whose *raison d'être* is the preservation of the state, cannot be subjected to the rule of law but instead allowed to suspend the law in the name of exception. In Schmitt's words, the rationale behind the state of exception – which he characterizes broadly as 'a case of extreme peril, a danger to the existence of the state, or the like' (Schmitt 1985, p.6) – resides in the premise that 'there is no norm that is applicable to chaos. For a legal order to make sense, a normal situation must exist, and he is sovereign who definitely decides whether this normal situation actually exists' (Schmitt 1985, p.13).

Schmitt's view of the state of exception as a *sine qua non* of sovereign power is of paramount importance in appreciating the role of frontiers in the constitution of the state and other forms of political rule. Schmitt himself later examined this role in the context of Europe's appropriation of the New World, a process that according to the author consisted of a series of bordering practices through which the Americas were juridically delimited as a 'free space' within which 'force could be used freely and ruthlessly' (Schmitt 2006 [1950]).[10] Schmitt's description of this process sharply captures the way in which the New World was built as vast frontier space, and how this space was instrumental in the making of a global (European) imperial order centred on the secular sovereignty of territorial states. Still, from Schmitt's perspective, the frontier is seen as a transient moment in the historical development of Europe's state system, just as the state of exception is justified as an imperative yet contingent means to protect the integrity of the state. It is through Giorgio Agamben's reconceptualization of this concept that we can come to understand the frontier as an immanent – rather than a spatially and historically bounded – condition of sovereign power.

Drawing upon Walter Benjamin's dictum that the state of exception has turned into the rule (Benjamin 1969, p.257), Agamben argues that the essence of exception is not that it designates a geographical or juridical space external to law but that it constitutes a relation that lies at its very heart and thus cannot be dissociated from it. In this sense, he points out that 'the exception does not subtract itself from the rule; rather, the rule, suspending itself, gives rise to the exception and, maintaining itself in relation to the exception, first constitutes itself as a rule', to which he adds that 'we shall give the name of *relation of exception* to the extreme form of relation by which something is included solely through its exclusion' (Agamben 1998, p.18). Put differently, in Agamben's reading of sovereignty, what characterizes the exception is not the act of juridical designation and suppression of 'chaos' (Schmitt) nor the 'state of nature' that precedes civil society (e.g. Hobbes), but a relationship of inclusive exclusion through which state power is constituted and preserved.

Agamben's contention that the state of exception constitutes a paradigm of government rather than a contingency measure allows an understanding of frontiers as spaces lying at the core rather than the periphery of the state order. This centrality, however, requires conceiving power in a topological rather than a topographical way – that is, not in terms of location and distance but in the spatial overlaps and porous borders between inclusion and exclusion or inside and outside (Allen 2011; Harvey 2012). The inclusive-exclusive relationship between state and frontier (the act by which the former subjects the latter by situating it in a relation of exteriority to law and order) is a clear example of a power topology that operates by establishing margins and borders that simultaneously include and exclude, or, in Agamben's own terms, by defining a 'threshold, or a zone of indifference, where inside and outside do not exclude each other but rather blur with each other' (Agamben 2005, p.23).

The bond between sovereignty and violence is firmly grounded in this topological relation of inclusive exclusion. This is so because as long as sovereign power resides in the permanent – and inalienable – capacity to suspend law in the name of exception, the preservation of 'chaos' (regardless of its temporal, spatial or political expression and its different incarnations: barbarian, primitive, savage, outcast, etc.) and its placing in a relation of opposition to 'order' is fundamental in every sense. Violence is exercised and legitimized through this relation of opposition, and remains unsanctioned as long as this relationship is maintained.[11]

There are plenty of instances of this (sovereign) violence in the spatial history of Colombia's Amazon frontier, many of which are rooted in this relationship of opposition and evidence how the frontier and the state have been constructed as two antagonistic yet indivisible orders: antagonistic, as they have been built up through a series of binary constructions ('civilization' vs. 'savagery', 'order' vs. 'chaos', 'Andes' vs. 'selva', 'White' vs. 'Indian' and so forth); and indivisible, for these same constructions have, since their inception, been mutually dependent and reinforcing. Put differently, this constitutes a relationship of opposition that has to be perpetuated, for it is through this opposition that the *illusion* of legitimacy of the state is sustained.

Rethinking the state and the frontier

The view of frontiers as spaces underpinning political control and violence has been variously formulated in the Colombian context.[12] Among this literature, the most systematic and exhaustive attempt so far to relate the production of frontiers with the origins and historical trajectory of the nation state is found in Margarita Serje's (2011) *El revés de la nación. Territorios salvajes, fronteras y tierras de nadie* [*The reverse of the nation. Wild territories, frontiers, and no man's lands*]. Serje's work constitutes a far-reaching journey throughout the multiple metaphors and discursive constructions through which territorial peripheries, margins or frontiers have been crafted in time and space, as well as the vital role these constructions have played in the consolidation of a hegemonic project of nation state.

This journey, which the author describes as 'an ethnography of the production of context', encompasses a wide array of characters and representational forms, from nineteenth-century foreign travellers and *criollo* elites' narratives and visions of the country's geography, to contemporary academics' discourses on the 'fragmented' character of the nation state in its various expressions, to official and non-governmental old and new recipes for 'development', and to NGO's and *hippies*' essentialist views on the preservation of 'nature' and 'culture'. The notion of context is of particular relevance, as it illustrates the process through which these narratives and views have become entrenched or established, thus determining 'a particular way of reading and interpreting reality as well as the ways in which it is possible to intervene upon it' (Serje 2011, p.37).

Serje's work constitutes a valuable effort to critically interrogate the historical and historiographical silences, erasures and misrepresentations through which frontier spaces have been discursively constructed, along with the continuous violence this process has entailed. *Frontier Road* is also concerned with the role of frontiers in state-building discourses and practices, and shares the view that state power is intimately linked to the preservation of different sorts of margins and borders. My focus on infrastructure, however, seeks to emphasize the importance of investigating not only the discursive but the material dimensions and everyday workings of power, an aspect that is absent in Serje's view of the frontier. In this sense, this book departs from, and aims to question, the view of such spaces as abstract constructions whose reality is solely confined to the realm of representation.

As I previously observed, there is little question about the violence that hegemonic constructions of the world have on people's lives and the spaces they inhabit. However, I argue that any attempt to unveil or historicize the genealogy of such constructions, must deal not only with their discursive or rhetorical dimensions but also with the localized contexts and material forms in which they originate and develop. Such an attempt will reveal that margins, peripheries or frontiers are not a passive locus of sovereign power (or, conversely, resistance) but *concrete* spaces where the power and knowledge practices of the state, capital or development are unevenly manifested and variously contested (Das and Poole 2004).[13]

In describing how the spatial history of the frontier has been shaped by a relationship of inclusive exclusion, I lay stress on the asymmetrical and violent nature of this relationship. Still, the very notion of relationship, uneven as it might be, implies interaction, which in other words means that margins or frontiers are not mere discursive projections of state or amorphous amalgamations of landscapes and peoples subjected to domination or, contrarily, sites of state resistance or avoidance. While this seems an obvious point, such notions of the frontier are commonly held and usually stem from the habit of seeing the state as an abstract force detached from or standing above society and nature. I would like to question this view by suggesting that state-building processes can only be fully comprehended if we take into consideration their discursive and material dimensions and, more crucially, the ways in which they are connected and mutually produced.

There are two related corollaries that stem from this assumption that are central to my argument on infrastructure. The first is that any attempt to approach the state ethnographically (e.g. by deconstructing and mapping the layers and practices through which it is configured, performed, contested and subverted) will find that it is far from a homogeneous and monolithic structure. This point has been particularly highlighted in anthropological literature, which has cast light on the relationships and interactions between the state and society, community, and culture.[14] Paying close attention to these relationships and interactions, as this literature suggests, represents a central task in studying the state, for they constitute an inherent – rather than incidental – aspect of state-making.

The second corollary is that these relationships and interactions cannot obscure the ways in which the power and agency of the state depends on its image as a self-contained and autonomous entity. In other words, as argued by Timothy Mitchell (2006), the task of studying the state implies not only refusing to take for granted binary constructions of political and social reality but accounting for *why* and *how* these constructions are produced. Regardless of the way we conceive the state (an 'instrument' of class domination, the 'monopoly of violence' on a given territory, an 'effect' of governmental or power technologies) this dual nature is essential to grasp the way in which it is crafted and manifests itself in practice. The main reason is that, in order to understand how power operates and is maintained, we have to account for the layers (material and discursive, symbolic and physical, concrete and abstract) through which it is produced and maintained.

Infrastructure, and roads in particular, provide a powerful means to examine state-building processes through those layers and the imbrications between them. At a very basic level, roads are physical structures that shape space in different ways, by enabling (and sometimes hindering) movement, settlement and control. Quite often, moreover, roads are part of larger policies and plans, from colonization schemes to the establishment of trade networks and the policing of territories. In this sense, they are structures that involve multiple actors and conflicts, and embody bureaucratic, ideological and political practices. Roads are built through engineering as much as they are built through such practices, and in this way not only constitute state technologies that shape or reshape space but configure spaces where the layers of the state are made visible.

Map of the book

In writing a spatial history of the Colombian Amazon that attends to such layers and their connections, I have sought to attend to the localized and concrete effects of power without losing sight of the larger power structures and processes at play. Thus, in retracing the history of the road, my central purpose has not been to build a chronological narrative of this infrastructure, but rather to situate its different characters, conflicts and events within the wider, long-duration process of state and frontier-making in the Amazon.

The first part of the book delves into the origins and consolidation of this process by narrating how the road was conceived and built, a story that begins with the early-nineteenth-century post-independence quest for geographical integration, and culminates in the early 1930s with the conclusion of the 230 kilometre bridle path connecting the Andean city of Pasto with the port town of Puerto Asís. This part draws extensively on government and missionary reports, travel narratives, cartographic representations, photographs and other archive sources in Bogotá, Putumayo and Barcelona. These documentary sources, which together constitute a practice of state-building, shed important light on the creative destructive process through which the Amazon was discursively and physically constructed as a frontier space.

Chapter One looks at the colonial genealogy of this process, as reflected through the rhetorical construction of the Andes *cordillera* as a physical and symbolic barrier separating 'civilization' from 'savagery'. The preservation of this image in nineteenth-century historical, geographical and cartographical representations of Colombia, constitutes a central background against which the vision of roads as powerful civilizing infrastructures emerged. In discussing how this vision became dominant, this chapter examines the exemplary figure of Rafael Reyes, a central character in the history of the Amazon and Colombia. The anatomy of Reyes in his different roles of entrepreneur, explorer and president, as well as a pioneer character in the history of the road, serves to reveal the violent ways through which the Amazon region was incorporated into the imaginary and spatial order of Colombia's nation state.

Chapters Two and Three tell the story of how this vision was put into practice. This is largely a story of struggle and violence amongst humans and between humans and nature that involves statesmen, Indians, missionaries, engineers, workers, *colonos* and other characters directly or indirectly engaged in the colossal project of opening a route across the rugged topography of the Andean-Amazon region. I place special emphasis on the relationship between the symbolic violence implied in the civilization/savagery dichotomy and the different forms of physical violence that this dichotomy sustained: the opening of the road 'breaking' the Andes through human labour and dynamite, the harsh political disputes over its control, the rampant grabbing of indigenous lands, and the persistent manifestations of confinement and abandonment from the *colonos* who worked in the road or arrived through it. The road's quotidian conflicts and dramas, together with the larger dynamics this infrastructure assisted or supported, fully epitomizes the ways in which the Amazon was assimilated into the order of the state as a frontier space. In discussing the rituals and practices through which this order was crafted and reinforced, I reflect on the notion of hegemony, and particularly on how it allows understanding of the everyday workings of state power.

From the early history of the road, the second part of the book turns to an ethnographic exploration of some of the instances in which the frontier–state relationship manifests in the daily life of the frontier, and of the different responses it elicits. Although these instances are diverse, they all draw attention to how transport infrastructures affect and shape people's lives in numerous ways. Chapter Four, for instance, reflects on conversations with different people (a local historian, a truck driver and a road activist), whose narratives bring to the fore the affective and lived realities that the *trampoline of death* provokes, from the hazardous practice of driving through its fragile and precipitous topography to enduring sentiments and memories of isolation, death, abandonment and fear. At the same time, however, these narratives show how frontier peoples make sense of and call into question their relationship with the state in spatial, historical and moral terms.

Chapter Five explores the question of state legibility in the context of a controversial road megaproject aimed at replacing the *trampoline of death*. The passage of the new road through an area of forests rich in biodiversity has been a point of contention on environmental and social grounds. Moreover, while the project has been promoted nationally and regionally as a prime example of sustainable development, the many conflicts and obstacles it has faced reveal the widening gap between its goals and outcomes. This gap was particularly evidenced in the policies and practices aimed at clarifying the complex land tenure situation in the project's area of influence. These policies and practices exposed the forms of knowledge and expertise through which this area was turned into an object of government intervention. However, they not only failed in bringing legibility to the area but actually made it more illegible. Through a detailed account of this process, I show how this 'illegibility effect' was produced, and how it generated multiple interactions and conflicts between state authorities, project officers and local communities.

Finally, Chapter Six focuses on the turbulent resettlement process of a community of forcibly displaced people illegally occupying a section of the road project's area. Through an ethnography of this community, this chapter investigates the political practices through which displaced peoples struggle for their rights, from the 'pirating' of public services and strategies to avoid eviction, to the everyday disputes and negotiations with local politicians and state institutions. I emphasize the unstable and often violent character of such practices in order to draw attention to the potentials and limits of what I call 'the politics of the displaced', as well as to highlight the exclusionary politics through which displaced peoples are included into the order of the state.

From the nineteenth-century utopian plans to civilize the Amazon through the building of waterways and road networks to the everyday conflicts and practices related to the current road megaproject, *Frontier Road* shows how a frontier was made and how it has remained. As noted, this is a story that involves many characters and events. Some of them encompass the entire Colombian Amazon and beyond, others the region of Putumayo – where the road is located – while others are confined to the physical space of this infrastructure or even fragments of it. All, however, relate to a territory that is relatively well defined in historical and geographical terms. Nevertheless, in reflecting upon these events and characters, I argue that the real meaning of frontiers transcends a specific spatial, temporal, or social context, and rather speaks of a *condition* of inclusive exclusion, regardless of the ways or forms in which it is expressed and materialized. I conclude this book by posing the road in parallel to other situations that affirm the violent effects of this condition, and render visible the borders and margins through which it is sustained in time and through space.

Notes

1   The term *páramo* refers to a grassland ecosystem located mainly in the upper parts of the northern Andes, in altitudes generally ranging from 3,000 to 4,500 metres above sea level.

2   This dimension has been especially addressed by ethnographic accounts of roads outside the so-called 'modern West'. See, for example: Campbell (2012); Columbijn (2002); Dalakoglou (2012); Harvey (2005); Kirskey & van Bilsen (2002); Lye (2005); Nishizaki (2008); Pandya (2002); Pina-Cabral (1987); Roseman (1996); and Thomas (2002).

3   See, for example: Campbell and Hetherington (2014); Harvey (2014); Harvey and Knox (2012); and Kernaghan (2012).

4   A clear example of this view can be found in Frederick Jackson Turner's classic account of the American frontier, where he defined the development of transport networks westward as 'lines of civilization' and their frantic expansion as 'the steady growth of a complex nervous system for the originally simple, inert continent' (Turner 2008, p.22).

5   This view was particularly prevalent from the 1940s to the 1970s among geographers and frontier historians influenced by Turner, Bowman, Bolton and other classical frontier theorists, who often portrayed roads as vital technologies to advance and develop the frontier. See, for example: Aiton (1994); Brücher (1968, 1970); Crist & Guhl (1947); Crist & Nissley (1973); Hegen (1963, 1966); James (1941); Townsend (1977); and Wesche (1974).

6   Since the 1970s, many scholars began to critically examine frontier processes, recurrently criticizing and denouncing their violent character. Dispossession of indigenous lands, environmental destruction, uncontrolled resource extraction and social conflict constitute some of the interwoven dynamics most commonly cited in this literature to describe such processes (see, for instance: Duncan & Markoff 1978; Foweraker 1981; Schmink 1982; and Schmink & Hood 1984). In the context of the Colombian Amazon see: Ciro (2009); Domínguez (1984, 2005); Fajardo (1996); Gomez (2011); Ortiz (1984); and Pineda (1987, 2003). A number of studies and monographs have specifically addressed the social conflicts and environmental impacts associated with road building, though most of this literature has focused primarily on the Brazilian Amazon (e.g. Fernside 2007; Moran 1981; Nepstad *et al.* 2001; Oliviera & de Moura 2014; Oliviera *et al.* 2005; Perz 2014; and Stewart 1994).

7   For a general historiographical review of the concept of frontier see Londoño (2003); Weber (1986) and Weber and Rausch (1994). In the context of Colombia see García (2003); Polo (2010); and Rausch (2003).

8   See, among others: Bolívar (1999, 2003); Bushnell (1993); García *et al.* (2011); González (1977); González, Bolívar and Vásquez (2003); Guhl (1991); Palacios (2007); Pécaut (1987, 2003); Safford & Palacios (2002); and Uribe (2001, pp.271–294). This argument has also been used to explain political fragmentation and interregional conflicts following independence and throughout the nineteenth century (e.g. Jaramillo 1984; Palacios 1980; and Park 1985).

9   Such influence has been emphasized in history and geography textbooks alike, as well as historical monographs. Among others, see Guhl (1976); Legrand (1986); McFarlane (1993); Melo (1986); Rausch (1999); and Reichel-Dolmatoff (1965). For a similar approach discussing the effects of Colombia's geography on economic development see: Gallup, Gaviria and Lora (2003); and Montenegro (2006). The costs and technical difficulties imposed by Colombia's topography on the building of railroads and roads constitutes a problem that has also been repeatedly emphasized by transport historians and scholars (e.g. James 1923; Morales 1997; Pachón & Ramírez 2006; Rippy 1943; Safford 2010; and Salazar 2000).

10 More recently, Schmitt's concept of exception has been taken up by a growing number of scholars interested in examining the relationship exception and political rule in a wide array of historical and geographical contexts. See, among others: Belcher *et al.* (2008); Legg & Vasudevan (2011); Mbembe (2003); Minca (2007); and Minca and Vaughn-Williams (2012).

11 Although Agamben's genealogy of the state of exception has been rightly criticized for neglecting the role in this process of colonialism and imperialism (Gregory 2006; Shenhav 2012), his theoretical and spatial approach to this concept has influenced a wide array of scholarship concerned with the subject of colonial and postcolonial sovereignty. This scholarship has critically examined the different ways in which exception has become constitutive to sovereign power in numerous historical and contemporary contexts and spaces, from colonial regimes and imperial projects to counter-subversion and anti-terrorism legislations, and to occupied territories, border crossings, refugee and migrant detentions camps, among others (see Svirsky and Bignall 2012 for a collection of works exploring the relevance of Agamben's work for colonial and postcolonial studies). Some authors, moreover, have theorized or explored the concept of frontier through the lens of exception, stressing how political sovereignty is often dependent on and realised through the production and maintenance of different types of margins or borders (e.g. Das Poole 2004; Hansen and Stepputat 2005; and Sundberg 2015).

12 For a discussion on the role of nineteenth-century imageries of frontier regions in the historical development of the nation state see Arias (2005); Múnera (2005) and Palacio (2006). In the context of the Amazon, the most thorough examination of the mimetic connection between colonial violence and representation can be found in Michael Taussig's *Shamanism, Colonialism, and the Wild Man. A study on terror and healing* (1991). Ramírez (2011) has explored the relationship between dominant representations of peasant communities and practices of state legitimation and violence in the context of the 1990s *cocalero* movements (in this same context see also Vásquez 2006). More recently, Wylie (2013) has examined the literary constructions of the Putumayo from the mid-nineteenth century to the present.

13   In the specific context of the Colombian Amazon there is a growing number of studies exploring and discussing local or endogenous processes of state-making, focusing mainly on actors, social conflicts and everyday political practices; see, particularly: Ramírez (2011); Ramírez *et al.* (2010); Torres (2007, 2011); and Zárate (2008). The issue of infrastructure, however, remains largely unexplored in the region, although it is becoming a topic of ethnographic interest in other parts of the Amazon (see, for example, Campbell 2012; Harvey 2012; and Harvey and Knox 2015).

14   For a general overview of this literature see, among others: Das and Poole (2004); Hansen and Stepputat (2001); Krohn-Hansen and Nustad (2005); Sharma and Gupta (2006); and Trouillot (2001). Other historical and ethnographic studies that have paid special attention to the social and cultural forms in which the state is embedded and reproduced include Carroll (2006); Coronil (1997); Kernaghan (2009); Nugent (1994); and Taussig (1997).

Part I

1

Reyes' dream

Is not the secret of the state, hidden because it is so obvious, to be found in space?

(Lefebvre 2009, p.228)

The December 30th sessions of the Second Pan-American Conference, hosted by Mexico from October 1901 to January 1902, were marked by a special event. On the date in question, General Rafael Reyes (see Figure 1.1), Colombian plenipotentiary to France and one of the country's delegates to the Conference, spoke of his explorations in the Amazon region in the 1870s, during which time he and his brothers were exporting quinine to Europe and North America. The presentation was not part of the ordinary Conference schedule, and despite the repeated insistence of his colleagues to 'reveal' his discoveries, Reyes, we are told, fearing 'he might be suspected of seeking notoriety by drawing public attention to his own person', was reluctant to break his 'modest silence' (Reyes 1979, p.5). Surely this gesture was more about a gentleman's etiquette, for the General not only jealously treasured his expedition notes but did not miss a chance to entertain his colleagues in private with his stories. No doubt he had repeatedly referred to his recent encounter in New York with President Theodore Roosevelt, to whom he gave an account of his journeys and presented his ambitious navigation project of the Amazon and its main tributaries. Mr Roosevelt, the eulogistic chronicler tells us, after enthusiastically listening to Reyes' account of the immense territory 'revealed' by him and his brothers, uttered the following words: 'That region is a New World undoubtedly, destined to

Frontier Road: Power, History, and the Everyday State in the Colombian Amazon,
First Edition. Simón Uribe.
© 2017 John Wiley & Sons Ltd. Published 2017 by John Wiley & Sons Ltd.

Figure 1.1   Rafael Reyes 1913.
Source: Library of Congress.

promote the progress and welfare of humanity' (Reyes 1979, p.5). Supposedly, following this encouraging encounter, Roosevelt had personally recommended that the US representative at the Conference use his 'best influence' in order to persuade the other delegates to give special consideration to his project. As for the Colombian General, with his discoveries having been praised by Roosevelt in such terms, he now felt it was a 'moral duty' to share them with his colleagues in Congress.

It was in this context that Reyes finally addressed the members of the Conference on the 30th December. His speech certainly must have captivated the audience, as he narrated his travels as a truly epic journey, where the terrifying presence of the unknown and the terrible privations and dangers endured by the discoverer were only surpassed by the incommensurable riches he unveiled and conquered for the sake of progress and civilization.[1] The opening episodes, those describing the crossing and descent of the south-eastern Colombian Andes in search of the Putumayo River, cannot but resemble the dramatic Spanish expeditions in the hunt for the elusive *El Dorado*.

> We started from the city of Pasto, situated on the summit of the Andes, under the equinoctial line. The immense region which extends from that city for more than 4000 miles to the Atlantic, was then *completely unknown*. We traversed a-foot the great mass of the Cordillera of the

Andes, which rises more than 12,000 feet above the level of the sea, up to the region of perpetual snow. Where this ceases there are immense plains, called paramos, upon which there grow neither trees nor flowers and where animal life completely disappears. We wandered for a whole month on those cold solitudes, guided only by the compass. They are covered with a fog as dense as that of the high latitudes of the North in winter; there were days in which we had to remain on the same spot in semi-darkness, without being able to advance a single step, the thermometer falling to 10 degrees below zero, a temperature made unbearable by the lack of shelter and shoes ... After marching for a month through that desert, in which perished, due to the intense cold, two men of the expedition, of the ten who carried provisions on their backs, we reached the limits of those solitary pampas which appeared like the product of a nature in progress of formation. We were at the Eastern watershed of the Andes. An ocean of light and verdure appeared before our eyes, in marked contrast to the shadows and solitudes which we had just traversed ... We penetrated these *unknown forests*, opening roads with the machete through brambles, briars and creepers which obstructed our passage. Arriving at the vertical slopes of the Cordillera, in places which were impassable, we had to descend by the aid of ropes (Reyes 1979, pp.14–15, emphasis added).

This was just the beginning. The 'Colombian Stanley',[2] as Reyes was referred to by the chronicler, continues his enthralling account by describing the sufferings and perils he and his brothers endured in their passage through the 'virgin forests', and then during their navigation of the Putumayo up to its intersection with the Amazon River. Along the Putumayo River they encountered numerous 'cannibal tribes', among them the 'powerful and warlike' Mirañas, of whom the daring General stated that he was 'the first white man whom those savages had seen' (Reyes 1979, p.16). The brothers made friends with the powerful chief 'Chua', who kindly offered them 'their dishes of human flesh' from their bitter enemies the Huitotos, and also provided them with oarsmen and canoes to continue their trip. After 15 days hunting and fishing with the Mirañas, they resumed their journey in company of the robust crew offered by Chua. It took them three months to descend the Putumayo River, a time which seemed to the brothers to be 'an eternity'. During the day they were exhausted by the extreme heat, the scarce food, and the fatigue of managing the canoe; at night, incessantly harassed by the dense clouds of mosquitoes, having to bury themselves under the burning sands of the deserted beaches along the river in order to avoid them. According to Reyes, they suffered the same fatigues as those endured by their 'savage companions'. Still, it was thanks to this circumstance, he stated, that they earned the affection and respect of the savages, 'who recognize no other superiority than that of strength' (Reyes 1979, p.17).

Their arrival at the Brazilian town of San Antonio, at the junction of the Putumayo and Amazon Rivers, marks a turning point in the narrative. The expedition had finally entered 'civilized' land again, six months after leaving the city of Pasto. The brothers had succeeded in their 'patriotic' enterprise of being the first in 'discovering' a river apt for the navigation of steamers, which would allow communication between the Colombian Andes and the Atlantic Ocean in Brazil. From San Antonio they caught a steamer to the city of Belém at the mouth of the Amazon, and from there they sailed for Rio de Janeiro. In Rio, as the news of their journey spread in the city, they were the object of numerous manifestations of applause and congratulation on the part of the authorities and distinguished personalities. The most prominent of them was the Emperor Dom Pedro II, whose 'majestic and commanding stature' and 'highly cultivated intellect' particularly impressed the young Reyes, who was then 25 years old. The Emperor, Reyes tells us, 'passionate for Geography and the exploration of the immense territories of his empire' (Reyes 1979, p.19), not only received him one afternoon in his palace, but listened with great interest to the account of his odyssey.

After two months in the capital, the celebrated explorers began their return journey to Colombia. They bought a steamer in Belém, which they navigated 1,800 miles upstream to the mouth of the Putumayo and then another 1,200 to its final stop at *La Sofía*, a river port that Reyes had named after his beloved fiancée. The climactic point of the journey, the moment that the small vessel made its triumphant entry into the waters of the Putumayo, was narrated by Reyes as a grandiose conquest:

> We can say that it was one of the happiest days of our lives, when we saw, for the first time, the flag of Colombia float from the stern of the vessel waiving in the breeze. This vessel was to realize the conquest of civilization and progress for our country and to improve the horrible condition of thousands of savages who at the mere contact with the civilized man felt as if struck by the electric spark of that same civilization, as they had not only treated us hospitably but very generously (Reyes 1979, pp.19–20).

At this point, the speech takes a radical turn. The arrogant and pompous voice of the conqueror now gives way to the sober tone of the statesman, who enlightens the audience with the country's inexhaustible resources waiting to be harvested by civilized hands: innumerable agricultural products; abundant gold, silver and emeralds; and thousands of 'savage Indians' that could be easily 'attracted to civilization', hence transformed into productive labour for the future enterprises established there. The exhaustive report, adorned with statistics and promising ventures, closes with the General's Faustian project: a colossal navigation system connecting the South American republics of Brazil, Bolivia, Peru, Colombia,

Venezuela and the Guianas through the immense waterway formed by the Amazon and its tributaries. This system was to link with another grand scheme – the Inter-Continental railway running from New York to Buenos Aires – hence allowing the aforesaid countries and 'humanity in general' to take advantage of 'the 4,000,000 square miles which the Amazon region contains and which it may be said is in its entirety uncultivated and uninhabited and consequently merely vacant land' (Reyes 1979, p.30).

He and his brothers, Reyes concluded, had significantly contributed to this enterprise. They had explored the Amazon and many of its tributaries, 'discovered' and established steam navigation in some of them, 'civilized' the 'savage cannibals' that 'formerly wandered' in the forests, and built trails linking the Putumayo lowlands to the Andes mountains. Sadly, the 'conquests' they had won for the 'progress and civilization' of their mother country and humanity, the General announced to the public, took a dire toll. During the years they spent in the rivers and jungles of the Putumayo, two of his brothers perished. Enrique succumbed to 'malignant fever', while Nestor was 'devoured by the cannibals of the Putumayo'.

When Reyes had concluded his passionate speech the reporter declared, in flamboyant rhetoric, that the audience was:

> galvanized with surprise by what they had listened to, with the delight of an exquisite satisfaction, by the contemplation of the very gorgeous pano-rama which the inspired narrator had unveiled before their sight, just as a magician exhibits before his public a series of enchanted palaces and gar-dens peopled by fairies and legendary genii (Reyes 1979, p.6).

A commission was appointed to verify the veracity of Reyes' account, and following its positive verdict the delegates unanimously made an appeal for 'collective action'. The Colombian explorer would be the object of numerous manifestations of gratitude, and his work, acclaimed as the 'base of a new geography', would be published in a single volume in Spanish, English, German, and French. The Assembly agreed to schedule a meeting to discuss his proposal at the next Conference (to take place in Rio de Janeiro), and issued a plaque in honour of the deceased brothers with the inscription: 'In memory of Nestor and Enrique who died in the service of the civilization of America' (Reyes 1979, p.11).

* * *

Reyes, elected president of Colombia in December 1903, never attended the Rio Conference. Furthermore, it seems that the General's project was not accepted unanimously by the delegates, and it would have raised 'disgusts and rancours' regarding its tacit acquiescence with US imperialist interests in the Continent (Marichal 2002, p.59). The project, however, would be resuscitated sporadically on future occasions, although to date the initiatives have largely remained on paper. Its most recent revival has been under the Initiative for Integration of Regional Infrastructure of South America (IIRSA), a massive continental initiative launched in 2000, which contemplates infrastructure development at multiple levels (transport, energy, telecommunications) aimed to enhance the physical and economic integration between the region's countries. In the Amazon, one of IIRSA's nine strategic areas, 64 infrastructure projects (57 of which are focused on roads and waterways) in eight different strategic corridors were originally projected, totalling an investment of nearly US$6 billion (IIRSA 2011). Among these projects is the development of a multi-nodal transportation scheme, whose chief purpose is to connect the Atlantic and Pacific oceans through Brazil and Colombia. The scheme's main components – the improvement of navigation along the Amazon and Putumayo Rivers and the construction of a 45 kilometre road section connecting the towns of San Francisco and Mocoa in the department of Putumayo – vividly evoke Reyes' project.

The significance of this project, in both its original and current version, lies not so much in its intended economic or political goals but in how it epitomizes the process of state-building in the Colombian Amazon over the last two centuries. This chapter traces this process throughout the nineteenth and early twentieth centuries. As I will describe in the first section, this process is inseparable from the entrenched image of the Amazon as an empty and savage space, an image whose origins can be traced back to the early times of Spanish rule and broadly mirrors the ways in which colonial spaces and populations were assimilated and appropriated. However, and since the chapter's main objective is to show how this particular image became a central feature within the foundational myth of the postcolonial nation-state, the analysis centres on the referred period. Moreover, the ways in which it surfaced, as will be illustrated in the second section, are inexorably linked to the post-independence quest for the geographical integration of the country, a quest that finds its major correlation in the dismal picture of the newly born republic as a mosaic of isolated, empty and autarkic regions. Within this order of things, state-building would be conceived as a teleological process through which the state would gradually but inexorably expand and absorb hostile or stateless territories and populations. At the same time, however, the civilizing mission of the state was utterly dependent on the savage image of the frontier. In other words, and as we witness in Reyes' 'magician's act' performed at the Pan-American Conference, 'state' and 'frontier' became part of the same rhetorical construction, the former's aura of authority and legitimacy built upon the 'savagery' and 'barbarism' of the latter.

It is important to emphasize that a historical analysis of this rhetorical construction and the spatial order it produced necessarily involves a wide array of characters, practices and representational forms, some of which will be considered throughout the chapter. However, the analysis will revolve primarily around the figure of Rafael Reyes. The relevance this historical character has for the arguments pursued here is based upon several aspects, although there are three main reasons worth mentioning. The first one is related to the theoretical and methodological approach of this work, which contrary to traditional views of the state as an abstract construction, focuses on the multiple material and discursive practices in which it is embedded. Secondly, as will be argued in the third section, the many facets embodied by this single character are crucial in understanding the particular discourses and practices through which the Amazon was constituted as a frontier space. Finally, as the story of the road begins to a large degree with Reyes, this chapter constitutes in many ways a preamble without which it is hardly possible to grasp not only the story itself, but the broader historical and spatial context in which it has unfolded since its beginnings up to the present day.

Two frontiers

The vast region extending from the east of the southern Colombian Andes to the Pacific Ocean that Reyes depicted as 'completely unknown' and which roughly alluded to the Putumayo and Caquetá river basins, was far from being *terra incognita* by the time he first set foot there. At that time – the early 1870s – the territory where he and his brothers spent several years devoted to the extraction of cinchona bark had for a while been incorporated into the country's territorial jurisdiction. This territory was then part of the *Territorio del Caquetá*, an extensive province established in 1845, which covered most of the country's actual Amazon and Orinoquia regions and had as its capital the tiny settlement of Mocoa, seat of the *Prefecto* (prefect), a priest, and a few *blancos* (mestizo settlers) engaged in different extractive activities. The *Territorio del Caquetá* had been surveyed and mapped in 1857 by the Chorographic Commission led by the Italian engineer and geographer Agustín Codazzi. However, for most of the nineteenth century it remained largely neglected by the central government as it was deemed to be a peripheral region of little political and economic interest for the country. Furthermore, the dramatic description that Reyes made of his crossing of the *cordillera* certainly constituted a common source of distress and torment, not only for the nineteenth century Colombian 'pioneers' and the few government officers stuck in remote and isolated outposts, but also for their European predecessors. During most of the three centuries of colonial rule, although the region would be widely penetrated by missionaries, *encomenderos* (grantees of Indians under the system of *encomiendas*[3]) and colonial officers (not to mention the early expeditions in the search of *El Dorado*), the Andes always represented a major barrier for the Spanish *colonos*.[4]

The Spanish foundations between the sixteenth and eighteenth centuries in this region were largely restricted to Mission towns, and were characterized by their ephemeral and tenuous existence. The case of Mocoa is in many ways exemplary of the colonial process of occupation of the Amazon. Considered to be the earliest Spanish settlement in the Colombian Amazon, it was originally founded around 1557 on a small valley in the Andean Amazon *piedemonte*, and named after the indigenous group inhabiting the area. Apparently, the initial settlement soon disappeared, for it was re-founded in 1563, the year that was regarded subsequently as its official founding date.[5] The Mocoas and other indigenous groups inhabiting the surrounding area were soon subjected to the system of *encomiendas*. For instance, as early as 1582, the Augustin friar Jerónimo de Escobar mentions that Mocoa had currently 800 indigenous people divided into ten *encomiendas*. The same friar, however, noted that the city had a 'bleak future' since communication with the rest of the provincial government was very precarious, and for this reason it was 'practically isolated'. Apparently, during that same year Mocoa was destroyed by the Andanquí Indians, to be reconstructed only decades later by the Jesuit missionaries. By the last quarter of the seventeenth century the city was again in ruins, the number of tributary Indians having been reduced to 75, and of the initial ten *encomiendas* only two remained (Llanos and Pineda 1982, pp.19–20). During the eighteenth century Mocoa continued to be the target of attacks on the part of the Indians, and for this reason was abandoned and relocated on more than one occasion. The decline of the missionary work reached its peak in 1784, the year when the Franciscans abandoned their Mission towns. According to Llanos and Pineda (1982, p.33), the overall balance of the Franciscan missionary work in the region was negative, and 'the colonization through the missionary regime had failed'.

The failure of the colonial policy of occupation of the Amazon through Catholic Missions does not mean that the Missions left no impact on the natives. The most visible was probably the demographic decline caused by smallpox and other diseases brought by Europeans and, at a more general level, the violence embedded in the colonial crusade. For instance, the recurrent rebellions against the friars and the destruction of Mission towns have been frequently attributed to retaliations against the violent practices implemented to 'reduce' the Indians to 'civilized life' (Llanos and Pineda 1982, p.37). Still, the opportunities that the Amazon indigenous peoples had to escape or avoid contact with the *colonos* were considerably greater than for those inhabiting places under tighter control by the Spanish authorities. Certainly, not only the vastness and difficulties of access made the missionary work in the Amazon lowlands a truly titanic enterprise; those friars devoting their lives to wandering the Amazon forests chasing 'unfaithful Indians' often perished under the inclement climate or were killed in indigenous revolts. The hopeless description that Fray Jerónimo de Escobar makes of Mocoa in 1582 is telling in this regard:

> This town is next to the mountains, far away from the road, so that it is a great travail to enter. Said town of Agreda [Mocoa] is not growing. Instead, it scares people away. There is no way to communicate and with the gold being extracted there, which can reach twenty-three-carat-gold worth some ten thousand pesos annually, with this they live and have a priest and a clerk, everyone having a miserable life (Escobar 1582, cited by Ramírez 1996, p.129).

The difficulties of access, together with the resistance of the Indians, the unhealthy weather and the lack of economic support by the Crown, largely explains why the missionary action in the Amazon ended up being confined to the uppermost parts of the Putumayo and Caquetá basins (Gómez 2011). Mocoa, despite its multiple resettlements and changes in name, and unlike the more eastern colonial outposts in the Amazon lowlands, would endure after three centuries of Spanish rule. However, its physical location in the *piedemonte* – a transition zone between the Andes and Amazon regions – came to symbolize a frontier between 'civilization' and 'savagery', and the abrupt trails connecting these regions metaphorically pictured as dreadful paths isolating rather than linking the two. Few descriptions embody this image so faithfully as this literary depiction of the ancient Pasto-Mocoa trail written by the Capuchin friar Canet del Mar:

> A nearly insurmountable barrier of the highest of mountains separated this vast land from Colombia. If an adventurer or a zealous missionary decided to overcome the obstacles that nature had in store, it was with great sacrifice and sometimes even endangering his own life. The road that communicated these savage lands to civilization was the most original and horrendous thing one could ever imagine; one would say that some malignant spirit had delighted in distributing precipices and abysses in order to block the entry to this solitary place, where savagery was rampant (De Pinell and Del Mar 1924, p.19).

The Dantesque experience that the missionary makes of the descent of the *cordillera* never seems to have existed among the region's native inhabitants. María Clemencia Ramírez (1996) has used the term 'fluid frontier' in order to allude to the rich cultural and economic exchange that since pre-Columbian times has existed between the different indigenous groups inhabiting the Andes highlands, the *piedemonte*, and the *selva*. Unlike the Europeans, for whom the *piedemonte* always represented a physical and imaginary barrier, the author illustrates how, for the indigenous peoples, this region has historically served as an articulation zone between the highlands and the lowlands. Mocoa, rather than the isolated town portrayed by Fray Jerónimo de Escobar, constituted a central crossroads where most of the indigenous trails converged. Apart from the mentioned path from Mocoa to the city of Pasto (described by the Capuchin friar and which constituted the 'opening act' of Reyes' presentation at the Mexico Conference), there were three other main exchange routes that together comprised a complex exchange circuit connecting the Putumayo and Caquetá lowlands and *piedemonte* with the Andean region (see Figure 1.2).

The indigenous groups of the *piedemonte*, and particularly the Quechua-speaking Ingas, had been, since pre-Hispanic times, specialized tradesmen. Products of the *selva* and the *piedemonte* such as dried fish, feathers, alluvial gold, bushmeat and wood resins were traded in the highlands for salt, tools, dogs and cotton. Although this exchange continued to exist throughout the colonial period and well into the nineteenth century, it was altered and transformed in different ways with the arrival of the Spaniards.[6] The ancient paths were gradually integrated to the colonial and early-republican exchange networks, and the indigenous peoples widely used as *silleros* (human carriers) not only for food and other products but for the missionaries, *encomenderos*, merchants, and other 'white' travellers.[7] Still, this new order was subverted in different ways, for not only the traditional exchange persisted among the indigenous peoples but it also facilitated the establishment of alliances against the foreign conquerors. Moreover, the intricate system of trails and paths allowed the development of smuggling routes, a trade in which the indigenous from the *piedemonte* took an active part. Such was the case of the Pasto-Mocoa trail, whose closure was ordered in 1751 to suppress the illicit trade of clothes that the Portuguese introduced by river to the upper Putumayo, and which the Sibundoy Indians carried on their backs for about ten days from there to the city of Pasto.

Through the persistence of traditional forms of exchange, the indigenous 'fluid frontier' survived side by side with a colonial spatial order

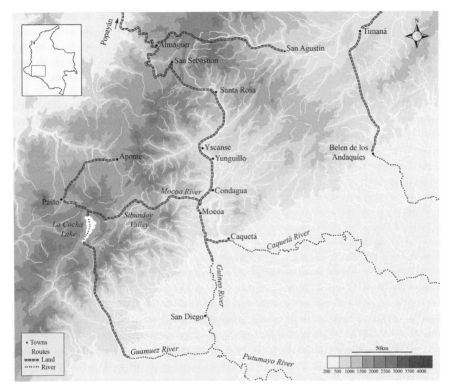

Figure 1.2 Indigenous exchange routes of the *piedemonte* c. XVI-XIX.
Source: Elaborated by author. Based on Ramírez (1996), Uribe (1995), Mapa de la Provincia del Putumayo siglo XVIII, AGN, Mapoteca 6, Ref.132.

in which the Andes and the Amazon appeared as two diametrically opposed worlds. This order of things would inevitably clash with the republican ideal of spatial and social integration in which the project of the nation state would be founded. Still, and paradoxically, as we shall see in the following section, it would be not in the annihilation of this order but on its own perpetuation that the power of the state was to be erected and sustained.

'The base of a new geography'

From the departure of the Catholic Missions in the last quarter of the seventeenth century until the creation of the *Territorio del Caquetá* in 1845, the extensive region known today as *Amazonia* (Colombian Amazon), remained practically isolated from the rest of the country.[8] Not only was this region considered of little political or economic interest, but the newly born republic concentrated its meagre fiscal resources in the most densely populated areas of the interior valleys and highlands. Even the interior or central provinces were largely isolated from each other, a situation which would persist throughout the whole century and redounded in the prevalent view of nineteenth-century Colombia as an 'archipelago' of a few populated centres separated by vast 'empty' territories (Melo 1986, p.151). The Magdalena River, which runs across the country from south to north and flows into the Atlantic, constituted the principal transport axis and the main export and import route. However, even after the introduction of steam transport around mid-century, the journey from the Caribbean coast to Bogotá could take up to a month of river navigation, plus another five or six days by foot or mule to cover the steep trail from the port city of Honda – where the navigation of the Magdalena was interrupted by rapids – to the capital. The Honda-Bogotá trail, although recurrently described by travellers to illustrate the arduous conditions of transport across the country, was in much better condition than the other routes connecting the Magdalena with the central and eastern provinces. Moreover, the development of transport infrastructure throughout the nineteenth century did little to ameliorate this situation. Whilst the road network hardly improved during this period, the boom of railroad construction since the 1870s essentially consisted of short and unconnected lines aimed at reaching the Magdalena's ports. This logic is largely explained by the fact that almost all of the railroads were designed to boost external trade rather than to enhance the precarious internal transport network. In addition, many were controlled by foreign companies and served to supply the industrialized world with raw materials, thus reflecting the outward-oriented nature of infrastructure development (Bushnell 1993, pp.134–135; Horna 1982; Safford 2010).

If this landscape constituted the 'civilized' side of the Republic, what would be the scenery of the vast Amazon region? The lament of the Prefect of the *Territorio del Caquetá* in 1850 is significant in this respect:

> Never will this territory escape from its ancient pitiful state, unless the difficulties are overcome and whatever possible is done to construct good ways in order to make the communication with the adjacent provinces possible (cited in Gómez 2011, p.64).

The Prefect's plea, as those from his future successors, would for several decades invariably end up filed in some dusty government archive in Bogotá. By the end of the century, the description of the Pasto-Mocoa trail by another Prefect showed how little the transport conditions had changed since colonial times:

> The journey from Pasto to this city [Mocoa] is gruelling, often bumping into horrifying places. Those of thin build travel on the back [of] Indians in a ridiculous, extravagant, and painful position: fastened with bale-rope and tied like pigs (cited in Gómez 2011, p.204).

The neglect of the Amazon region by the central government was also felt in the abandonment of the borders with neighbouring countries. Reports of an indigenous slave trade along the Caquetá and Putumayo rivers, a practice which dated back to the seventeenth century and was largely carried out by Portuguese merchants, were frequent throughout the nineteenth century (Llanos and Pineda 1982; Pineda 2003). However, the major repercussion of this neglect would be a series of long and intricate territorial disputes, mostly with Peru and Brazil, which resulted in the loss of an extensive strip of land between the Putumayo, Napo and Amazon rivers. The country's weak border policy over this century is partly attributed to the government's strategy of claiming its territorial rights on the basis of the *Uti possidetis iure* legal doctrine, or the principle through which the newly born republics would preserve the colonial limits at the time of independence.[9] However, not only were the boundaries between the former colonies very confusing in regions such as the Amazon, thus facilitating the *de facto* appropriation of territories in dispute, but the practical measures of successive Colombian governments to safeguard its borders were negligible throughout the nineteenth and well into the twentieth century, a situation mirroring the state's blind faith on the *Uti possidetis iure* principle (Palacio 2006, pp.133–142; Zárate 2008, p.188).

As early as the 1890s, when the extraction of rubber was just beginning to emerge in the Colombian Amazon, the consular agents in the Amazon cities of Iquitos, Manaus and Belém continually warned the government of the regular incursions into national territory by Peruvian and Brazilian *caucheros* (rubber tappers). The complaint of the Colombian Consul in Belém to the Minister of Foreign Relations in 1894 epitomizes the drama of the borders:

> From the three rivers Napo, Putumayo and Caquetá large quantities of rubber and many other natural products are currently being extracted enriching other countries ... this is so because, sadly, Colombia lacks there any presence of authorities able to guarantee her territorial domains. With all due respect, I cannot comprehend the attitude of this government which looks with such unreasonable indifference to such precious national interests.[10]

Against this grim picture of the *desiertos orientales* (eastern deserts) – as the Amazon frontier was often referred to in the nineteenth century – it comes as little surprise that this region integrated first to the world economy through successive extractive booms such as cinchona and rubber, rather than to the Colombian state. Even nowadays the older *Mocoanos* invoke memories of the rubber boom, around the dawn of the twentieth century, when the circulating currency was the sterling pound and European biscuits and pastries were available at the local market for those who could afford them. Not long afterwards, when Rafael Reyes had resigned from the presidency of Colombia and was exiled in Europe writing a memoir about his youth for his sons, he opened the chapter on the Putumayo with the following words:

> In Pasto the region that extends to the east was known only as far as Mocoa, and beyond there the populace, ignorant of geography, thought it was Portugal; they confused this country with Brazil. Those forests were populated by monsters and terrible beasts, something alike the unknown and fantastic world which must have been for humanity the seas and regions that Columbus discovered (Reyes 1986, p.109).

The revealing nature of Reyes' statement lies not so much in his judgement about the Colombians' rampant ignorance of the country's peripheral regions – which was by no means restricted to the 'populace' – but in how this ignorance was superseded by all kinds of imageries and tropes. Through these imageries and tropes, those peripheral spaces such as the *Territorio del Caquetá* would be discursively assimilated within the imaginary order of the nation long before they were to be physically integrated to the spatial order of the state. These two apparently dissociated processes, however, were inexorably connected, since the ways in which the latter was produced and normalized can only be understood by first addressing the logic behind the constitution of the former.

Creole pioneers

Benedict Anderson (2006) has used the expression 'creole pioneers' in order to account for the early rise of nationalism in Latin America, a process that was marked by the emergence and proliferation of independence movements across the continent since the late-eighteenth century. According to Anderson, the two factors commonly cited to explain the development of these movements – the spread of

Enlightenment ideas together with the tightening of Spanish control over its colonies during the second half of that century, although central in understanding their origins and evolution, do not themselves explain how they became 'emotionally plausible and politically viable' (Anderson 2006, pp.51–52). A more thorough explanation, suggests the author, must be sought in the articulation of two different yet related factors. First, the fact that the Spanish administrative units, whose original shape derived from arbitrary or fortuitous circumstances, developed over time as isolated and self-contained units. This self-contained character, which to a large extent resulted from the Spanish policies of turning administrative units into separate economic zones, was reinforced by geographical factors, in several cases translated into physical barriers and immense difficulties of communication between the colonies. This situation helped explain why the independent states were initially created according to the colonial territorial jurisdictions, as reflected by the adoption of the *Uti possidetis iure* principle.

However, according to Anderson, the autarkic nature of the colonial territories alone does not account for the sort of attachments that made possible the transition from those administrative units into independent nation states. The origins of those attachments and their materialization into nationalist movements have instead to be sought in the political and economic exclusion faced by the creole (*criollo*) society.[11] This exclusion, partly associated with the racial stigma of being born outside the Metropolis, entailed a crucial dilemma out of which the *criollos* found a common identity against the Spanish-born Spaniards: for if 'born in the Americas, he [the creole] could not be a true Spaniard; ergo, born in Spain, the *peninsular* could not be a true American' (Anderson 2006, p.58).

Anderson's argument on the 'creole pioneers' provides a good summary of the process through which, thanks to a series of interrelated geographical, economic and social factors, the Spanish territories in America gradually evolved into independent nation states or 'imagined communities'. However, this argument does not account for the ways in which those communities were actually *imagined,* and how through this new imaginary order we find not the culmination of colonial forms of domination and control but their own perpetuation. Mary Louise Pratt has clearly elucidated this point when she notes that 'politically and ideologically, the liberal *criollo* project involved founding an independent, decolonised American society and culture, while retaining European values and white supremacy' (Pratt 1992, p.125).

This is not the place to discuss the multiple and contradictory ways in which nationalism and state-building in Colombia or Latin America in general became deeply entangled with constructions of race and class, and how those constructions translated into myriad forms of political, social, economic and spatial exclusion.[12] There are, however, two aspects that I would like to stress as they are central to the argument presented here. First, we have to acknowledge the rootedness of those ideologies in the imaginary order of the nation in order to fully grasp the colonial logic in which this order was and is still embedded. Secondly, we need to look at how within this same order the idea of race became inexorably attached to space to understand the ways in which the post-independence ideal of social and spatial integration was conceived and put into practice.

One of the best illustrations of how this *criollo* imaginary order was conceived is the early assessment of the country's geographical knowledge made by the independence martyr Francisco José de Caldas. In an article published in 1808 in the *Semanario del Nuevo Reino de Granada* – the main scientific journal at the time - Caldas, also considered to be the father of Colombian geography, made an urgent appeal for the need to overcome the absolute ignorance regarding the country's geography. The self-taught geographer and astronomer summarized this state of affairs in the following words:

> Let's take our gaze to the north, let's take it to the south, let's register the most populous parts or the deserts of this Colony: everywhere we find nothing but the stamp of sloth and ignorance. Our rivers and mountains are unknown to us; we ignore the extension of the country in which we were born, and our geography is still in its cradle (Caldas 1966, p.208).

In the same writing, Caldas elaborated a basic classification of the country's population, which he broadly divided into 'savages' and 'civilized'. By the former he specifically meant the 'wandering' and 'barbarian' indigenous tribes inhabiting the vast peripheral forests, savannahs and deserts. The latter, comprising those living under the 'laws of society', were subdivided into three differentiated 'races': the 'civilized Indians'; the 'Africans' introduced as slaves after the discovery of the New World; and the 'European conquerors'. This last category included a further classification, as Caldas emphasized that by 'Europeans' he meant 'not only those who were born in that part of the world, but also their sons, who preserving the purity of their origin, have never mixed with the other castes' (Caldas 1966, p.188). This group explicitly referred to the *criollos*, the caste to which Caldas proudly belonged, and of which he stated they represented 'the nobility of the Continent'.

Caldas' hierarchical ordering of the country's inhabitants was not restricted to their racial origins, and to a large extent reproduced the prevailing environmental determinism that dominated Western thinking and underpinned nineteenth-century scientific racism. Caldas had elsewhere written about the influence of climate on the country's races, taking into consideration a wide range of variables such as atmospheric pressure, temperature, electric charge, wind, rain and altitude (Caldas 1966, pp.79–120). However, and despite the racial variances and moral virtues and vices stemming from these variables, a geographical division of the country in two broad (inhabited) zones prevailed throughout the text: the temperate regions (predominantly the populated areas of the Andean mountains), whose mild climate was directly associated with the 'industrious' and 'intelligent' character of their inhabitants; and the hot and humid regions of the *selvas* and coasts, the 'natural' habitat of the 'savages' and whose 'scorching' heat and excessive humidity condemned humans to a perpetual state of 'barbarity', 'laziness' and 'backwardness'.

This dichotomous order, whose origins hark back to colonial times and became pervasive within nineteenth century elites' construction of the nation (Arias 2005, pp.69–82), assumed over time different expressions such as 'highlands' versus 'lowlands' or *tierras frías* and *templadas* (cold and temperate lands) versus *tierras calientes* (hot lands). However, the fundamental significance of this order is not its perpetuation itself, but how it acquired a hegemonic character through which space became *racialized* or –conversely – race became *spatialized*. This hegemonic character, as noted by Wade (1989), lies not in the fact that it was uncontested, but in how its principles came to appear as self-evident truths, truths whose appearance of reality cannot be detached from the power relations in which they are sustained. It is in this sense that one can easily appreciate the role played by the *criollo* elite – among which Caldas figured as one of the most prominent characters – as a class of *organic intellectuals* in the Gramscian sense, or a class whose function is 'conceived as absolute and pre-eminent, and their historical existence and dignity abstractly rationalized' (Gramsci 1971, p.117).

Caldas utterly embodied this intellectual function, as he captured the pre-eminent role of science and particularly geography within the *criollo* state project.[13] He condensed the essence of this role in his widely quoted statement that 'geography is the fundamental basis of all political speculation' (Caldas 1966, p.183). The transcendence of this statement has to be appreciated not only in the epistemological and philosophical principles underlying geographical discourse, but chiefly in its *practical* implications and effects. For, if Caldas' hierarchical ordering of the country essentially opposed the 'civilized' environment of the *cordillera* to the untamed world of the *selvas* and savannahs, it also entailed a crucial paradox: not only the latter encompassed most of the country's territory but – and most significantly – they were imagined as an infinite container of natural resources waiting to be unveiled and harvested for the sake of 'progress' and 'civilization'. Who were meant to carry forward this nationalistic enterprise? Naturally not the hordes of 'savages' wandering in the jungles, whose state of 'barbarism' was only surpassed by their 'laziness' and total lack of entrepreneurship. Since the white races of the country – represented mostly by the *criollos* – comprised just a minority of its population, this task would ideally be accomplished with the aid of European immigration. Throughout the nineteenth century the central government unsuccessfully attempted, on various occasions, to encourage foreign immigration through laws that offered land grants and other incentives to potential *colonos*.[14]

The explanations of why the attempts invariably failed are varied and range from climate to politics and from geography to economics. However, it is important to mention that the reasons were not only confined to local factors such as the government financial constraints; 'global' causes such as nineteenth-century Europeans' fears regarding acclimatization in tropical areas also explain an important part of the picture (Uribe 2015). The point I would like to highlight is how through these initiatives and, overall, through the *criollo* elite's faith in the white race's 'enlightening' powers, the country's vast 'peripheral' regions were proclaimed as 'vacant' or 'empty' lands waiting to be possessed. Caldas' desperate call for a meticulous and thorough survey of the country has to be understood precisely in this sense, for the question was not only about the vital need to overcome the deplorable state in which the geographical knowledge of the country was plunged in as how, and for whom, this knowledge was to be constructed.

Caldas' dream would take decades to be realized, but it was finally achieved during the 1850s through the Chorographic Commission, a state-sponsored project that is considered the single most important geographical event of the century in Colombia. The Chorographic Commission, of which Caldas is regarded as the precursor as well as being responsible for conceiving its purposes and 'ideological foundations' (Sánchez 1998, p.69), consisted of a detailed survey and a geographical chart of each of the country's provinces. The contract signed between the government and the head of the Commission, the Italian military engineer and geographer Agustín Codazzi, clearly expressed the purpose of this colossal project: '[the descriptions and maps] must have the adequate extension, clarity, and precision, so that the country can be known and studied in all its dimensions, particularly in relation to topography, statistics and natural wealth' (cited in Sánchez 1998, p.239). Thus, the surveys and charts should gather an immense quantity of data that included inventories of vacant lands, natural resources and agricultural production; relations of existing trails and paths with distances and times of travel; location and commerce statistics of towns and villages; topographical descriptions of rivers, mountains, valleys and forests; and accounts of climate and populations. Not surprisingly, the issue of transport, a nightmare for travellers since colonial times, occupied a special place within the Commission's objectives. As noted by Sánchez (1998, p.238), the fact that Codazzi was appointed as 'roads engineer' instead of 'geographical engineer' largely mirrored the consideration given to this subject by the national government.

The precarious situation of the country's transport network, which the Chorographic Commission was expected to improve through the identification and projection of new routes, paradoxically constituted a considerable obstacle for the Commission itself, causing several delays in the works and torments to its members. Not unexpectedly, of the numerous expeditions carried out by Codazzi between 1850 and 1859 – the year of his death – none caused the Italian geographer so much suffering as the forsaken *Territorio del Caquetá*. In his letter to the Secretary of State notifying the conclusion of the expedition to Caquetá, the restless engineer wrote:

> I have happily left behind the Andaquíes [Caquetá] after having sketched the map of that extensive and unhealthy desert ... I can assure the government that none of my expeditions has cost me so much money, nor have I suffered that many torments, neither have I seen myself, as I have on this occasion, so often exposed to die (Codazzi 1996, p.237).

The *Caquetá* expedition, carried out between January and April 1857, was confined –mostly due to the difficulties of access – to the upper Putumayo and Caquetá basins, and for this reason Codazzi based his report mainly on a series of written sources.[15] The opening lines of his description of the territory strikingly mirror the *criollo* vision of the peripheral 'deserts' and 'selvas': 'None of the ancient provinces in which the Nueva Granada[16] was divided' – wrote Codazzi – 'can compare in dimension to the extensive Territorio del Caquetá; and yet, this territory is the most *deserted* and the *least inhabited* and known of the Republic' (Codazzi 1996, p.151, emphasis added). The rhetoric of the civilized world of the *cordillera* versus the savage and yet boundless natural wealth of the *selvas* is continually reproduced throughout the text, as can be seen from this fragment that is worth quoting in full:

> There is no space on the ground which is not covered like a carpet by a diversity of plants. In the midst of such magnificent vegetation in which man has not had the least part, he almost finds himself like an imperceptible being in the middle of that vast land where everything is enormous: hills, plains, rivers and jungles. Upon seeing the gigantic

development of the organic forces, of that overwhelming wealth, he
realises that a numerous population is required to dominate such por-
tentous vegetation. Time, a long time, is needed for man to be able to
exploit the immense wealth that the land offers in an incalculable pro-
fusion (Codazzi 1996, p.197).

The indigenous population, which Codazzi calculated as being
50,000 – most of them 'savages'– was to him clearly insufficient to
exploit the 'overwhelming' wealth of the *Territorio del Caquetá*, and
more so when he estimated that the whole territory – which comprised
roughly half of the country's territory – could easily 'contain' a population
of 23 million (see Figure 1.3). Moreover, like Caldas, he was highly pessimistic of the natives' agency and endeavour. This was especially the
case of the eastern parts of the territory, those which the colonial state
was never able to control – let alone 'civilize' – and which, Codazzi
judged, were still as 'backward' as the world Columbus encountered.
'The savages', he declared:

> Will make no progress until the *criollos* get into close contact with them:
> otherwise they will never escape the state of barbarism in which they are
> born, live, and die, without knowing anything other than the satisfaction
> of their most vital needs to live brutally, almost like the beasts of the forest
> (Codazzi 1996, p.194).

Yet, although Codazzi saw the climate as the main obstacle for national
and foreign white races' chances to take advantage of the abounding
natural resources of *Caquetá*, he showed an unfettered faith in the transformative power of capitalist development. Thus, as with the 'new geography' Reyes outlined decades later at the Pan-American Conference, the
region Codazzi projected was one dominated by transoceanic steamship
navigation, water channels, railways and roads. In this utopian infrastructural landscape, predicted Codazzi, the 'pestilent' climates of the *selvas*
would be thoroughly modified as soon as a 'numerous population' had
'cut down the old trees of the forest, drained the marshlands and swamps
and channelled the rivers' (Codazzi 1996, p.201).
Codazzi's description of the *Territorio del Caquetá* can hardly be
reduced to a mere reproduction of nineteenth-century racial and environmental determinist doctrines, and there is no question that his
account of the territory, which along with a detailed geographical
survey consisted of several maps and drawings as well as rich ethnographic and botanical descriptions, has immense historiographical
value. However, we have to locate this singular event in the broader
discursive apparatus – philosophical, scientific and political – through
which the *criollo* imaginary order of the nation emerged and gradually

Figure 1.3 *Chart of the Nueva Granada, divided into provinces, 1832 to 1856, Uti possidetis of 1810. Territorio del Caquetá* shown at the bottom of the map. Source: Codazzi, Paz and Pérez 1889.

acquired a hegemonic character. Foucault's idea of discourse as a 'field' where subjects are unevenly located or 'a space of differentiated subject-positions and subject-functions' (Foucault 1991, p.58) is enlightening in this respect, for what we see in the colossal project of the Chorographic Commission is not only the production of knowledge about the country, but its normalization into an already established discursive field.

I want to place emphasis on this process of normalization, since it is precisely here that we encounter one of the central axioms on which the foundational myth of the modern state is grounded. This myth, which harks back at least to the idea of the social contract that we find in Hobbes' *Leviathan*, stems from the philosophical fiction of the 'state of nature' as the legitimization of a supreme sovereign endowed with the power to impose security and peace among its subjects. Under the state of nature, Hobbes tells us, there is 'no Society; and which is worst of all, continual fear, and danger of violent death; and the life of man, solitary, poor, nasty, brutish, and short' (Hobbes 1937, pp.64–65). Reading

Hobbes' *Laws of Nature* one cannot but think of the *criollos'* chaotic vision of the country's jungles as a vast space infested with savages, whose existence is hardly differentiated from the 'beasts of the forest'. Once this vision was embedded into the landscape through the erection of a hierarchical and racialized spatial order, the state *appears* as a sovereign force whose legitimate existence is sustained on its civilizing character. State-making, accordingly, is seen as a teleological process through which hostile territories and populations are gradually integrated into the civilized order of the state. Yet, this idea finds a major paradox on the same principle on which it is founded. For, if the illusion of legitimacy in which the power of the state ultimately rests lies on the binary opposition between 'civilization' and 'savagery', does not this opposition need to be perpetuated so the illusion can be maintained? This actually constitutes an essential paradox that we encounter in the *criollo* project of the state: a project aimed at the social and spatial integration of the nation, and yet a project whose legitimacy is sustained on the perpetuation of the civilization/savagery rhetoric on which it is founded.

It is also in this context that the idea of the frontier as a metaphor designating those spaces lying beyond civilization emerges as a central element within the foundational myth of the state. As state and frontier come to embody the binary opposition between civilization and savagery, the paradox is thus maintained: the frontier seems to be inescapably destined to vanish as the state expands, and yet it cannot totally cease to exist, for without the frontier the myth in which the power of the state is founded would also vanish. Thus, although the frontier comes to appear as the antithesis to the civilized order of the state, its *status* of frontier inevitably becomes not a barrier to this order but its very condition of possibility. The frontier, however, will not come to encompass a space located outside the order of the state, but one that lies at the very core of this order. In other words, as previously noted, the frontier was to be constituted and demarcated as a space of exception in the Agambean sense – that is, a space resulting from the 'extreme form of relation by which something is included solely through its exclusion' (Agamben 1998, p.18). This relation is precisely what we find in the *Territorio del Caquetá*: a territory whose incorporation to the spatial and political order of the state has historically depended on its exclusion from the imaginary order of the nation. In Codazzi's judgement about this territory as a 'vacant space' waiting to be occupied and possessed, it is possible to foresee some of the practical implications underlying this form of relationship. However, as will be shown in the following section, it is through another character, Rafael Reyes, that both the myth and the immanent violence sustaining this relationship will surface.

The secret of the state

One of the main problems in building a theory of the state, as noted by Timothy Mitchell, is that the state constitutes an 'object of analysis that appears to exist simultaneously as material force and ideological construct. It seems both real and illusory' (Mitchell 2006, p.169). According to the author, most analyses tend to dismiss one dimension in favour of the other: either they take for granted the binary dualism through which the state seems to exist as an autonomous realm completely detached from society, thus assuming the state as an 'abstract' construction; or they reject this dualism as mere ideological fetishism, hence adopting a perspective that privileges the study of the multiple material relations and practices embedded in the state. Although Mitchell agrees with the latter perspective in the sense that any attempt to theorize the state cannot take for granted this dualism, he argues that it is not enough simply to criticize it. Such perspective, he adds, not only ignores that it is in this dual form that the state often appears in practice, but that the agency of the state largely depends on the production of this dualism. The task of critique, accordingly, is then not simply to reject the dualism but to explain *how* the effect through which state and society appear in this dual or binary form has been produced.

The relevance that a figure like Rafael Reyes has in understanding how the Amazon region was discursively and physically constituted as a frontier space has to be considered in this sense. Like Caldas or Codazzi, Reyes personified the *criollo* hegemonic vision of the frontier as the antithesis of civilization. However, Reyes' role was not exclusively circumscribed to the sphere of scientific discourse or political rhetoric. Through the different facets he embodied throughout his life – entrepreneur, explorer, army officer, diplomat, statesman – it is also possible to shed light on the material (spatial, infrastructural, political) practices of state-building. I will focus specifically on two of these facets, since it is through them that the relationship between the Amazon frontier and the Colombian state can be better grasped: the *entrepreneur* and *explorer*, referring to the years he and his brothers spent in the Putumayo engaged in the cinchona and rubber trade; and the *statesman*, covering roughly the period during which he was president of Colombia (1904–1909).

Finally, it is important to stress that the point here is not to assume Reyes as a single individual who supplanted the state or seized its roles, but as the expression of certain views and practices that are both constitutive and reflective of the relationship between state and frontier. In other words, as with Marx's capitalist, rather than an individual we are dealing with certain types of 'personifications' of particular relationships (Marx 1949, p.xix).

More specifically, I want to situate and encourage the reader to see Reyes in the broader context of the spatial history of state and frontier as a maker of history in the threefold dimension drawn by Trouillot: as an *agent* or individual part of a certain class or hierarchical structure; as an *actor*, whose role and actions are circumscribed to a specific spatio-temporal context; and as a *subject* or voice aware of his power in the production of certain historical narratives (Trouillot 1995, pp.22–24).

The pioneer

The project Reyes conceived during his early explorations in the Putumayo and which he made 'public' at the Pan-American Conference in 1901, would stay with him until the end of his life. After having resigned the presidency in 1909, the tireless General, now in his early 60s, devoted a few years to travelling across Europe, the United States and the South American republics, where he continued to promote his continental integration project and crusaded in favour of the Pan-American union. An account of these travels was published in Spanish and English in 1914 and reproduced in instalments in the *New York Times* (1914). The US newspaper, although recommending his work due to the 'official prestige of his author' and the 'intrinsic interest of the narrative', regretted the fact that the author did not describe in detail the story of his adventures in the 'thrilling no man's land' of Putumayo. Certainly, the chapter on the Putumayo basically consisted of an abridged version of the paper he presented in Mexico years before. However, Reyes added here a short introduction where, in a paragraph that evokes the opening episode of Joseph Conrad's *Heart of Darkness* – where Marlow recalls his childhood obsession with the blank spaces on the world map – he tells the reader about his early fascination for the 'unknown' Amazon forest. 'From my very childhood', he writes:

> I felt myself attracted by the mystery of those immense forests. I used to cherish plans for exploring them, and of opening across them a communication with the Atlantic, thus giving new channels for commerce and for the glory of my fatherland (Reyes 1914, pp.41–42).

Reyes never published a detailed account of the time he spent in the 'mysterious' Putumayo. However, he narrated this story in a series of notebooks and letters to his sons that he wrote during the early 1910s, which were published posthumously as his *Memoirs* (Reyes 1986). There, he tells how he ended up there, a story that seems to have more to do with chance than with premeditated resolution. As young as 17, Reyes begins, he had concluded that his native land – the small town of Santa Rosa de Viterbo, located in the then sovereign state of Boyacá – was 'too narrow a theatre' for his 'great ambitions' (Reyes 1986, p.32), and therefore considered travelling to Panama or California, a common destination at the time for fortune seekers. He was about to leave when his mother got a letter from his elder half-brother Elías in which he asked her to send Rafael and Enrique to assist him in an import business he had established in the city of Popayán (then the capital of the state of Cauca, to which the extensive *Territorio del Caquetá* belonged). As the latter decided to stay in order to take care of the family – Reyes' father had died several years previously – the young and ambitious Rafael, without hesitation, embarked on foot and by mule on the long and arduous journey to Popayán.

Not long after he had joined his brother Elías, Reyes saw a promising business opportunity in the exportation of cinchona tree bark, out of which quinine was extracted, a substance known for its antimalarial properties that was at the time in high demand in Europe and the United States. Since the cinchona trees grew in abundance in the southern Andean foothills and they remained largely unexploited, Reyes undertook a series of expeditions to buy the bark from the few 'whites' living there and also to explore new extraction areas. It was during one of those expeditions – across the slopes of the *cordillera* to the east of Popayán – that Reyes mentioned that he saw for the first time the immense Amazon lowlands. Watching from the top of a tree, a scene that vividly evokes a famous scene of Werner Herzog's movie *Fitzcarraldo*, he could not but marvel at the 'endless and immense green ocean'. Recalling his bewilderment with the grandeur of the Amazon, a theatre that at last seemed to be big enough for his 'great ambitions', he writes:

> Those virgin and *unknown* forests, those immense spaces, fascinated and attracted me to explore them, to traverse them and get to the sea, and to open roads for the progress and welfare of my country; those forests were *absolutely unknown* to the inhabitants of the *cordillera*, and the idea to penetrate them terrified me since the popular imagination populated them with wild beasts and monsters, besides the numerous savage cannibals found there (Reyes 1986, p.81, emphasis added).

This proclamation, that Reyes invariably uses whenever he introduces his expeditions, and which is revealing of the fusing of patriotism and self-ambition so typical of his character, fully reflects the *criollo* vision of the frontier. Yet, as it has already been noted, in Reyes we witness not only the vision but also the state practices through which the frontier was constituted. The commercial activities – mainly around the extraction and export of cinchona bark – he and his brothers undertook in the Putumayo during the 1870s constitute a remarkable example of both the vision and the practice. Reyes' best-known biography, for instance, not only states that the Reyes brothers' company (*Elías Reyes y Hermanos*) was the first large-scale commercial initiative founded in Colombia, but also eulogizes that:

> We don't know of any [company] which, under private initiative, *without official support* and with no political ambitions of any kind, had mobilised such quantity of men and money towards a *licit, and at the same time progressive and patriotic goal* (Lemaitre 1981, p.89, emphasis added).

Although this statement might appear exaggerated, it sharply captures the discourse with which Reyes himself infused all his commercial projects and achievements. Moreover, the significance that the Reyes brothers' company has in the history of the Colombian Amazon is widely recognized.[17] In this sense, as previously indicated, even though this territory was far from being unknown at the time Reyes had his great epiphany at the top of a tree, it is a fact that prior to he and his brothers starting their extraction activities, the presence of Colombians in the region – excluding the 'savage tribes' – was insignificant. This situation changed significantly in the following decades as a result of the cinchona and especially the rubber boom, the Reyes brothers clearly counting among the pioneers of that infamous episode in the history of the Amazon.[18] It is not accurate, however, that the brothers did not have any official support, as they obtained from the government a large land concession in the upper Putumayo and Caquetá basins, an area abounding in cinchona forests (Domínguez 2005, p.87). However, rather than being an unintended omission, the biographer's imprecision reflects the widespread view of the Amazon as a vacant frontier waiting to be grabbed.

From a commercial standpoint, the brothers' venture could be considered to be a truly remarkable story of endeavour and achievement. In 1875, as the cinchona forests to the east of Popayán began to be exhausted, they focused their activities in the *piedemonte*, and Mocoa became the Company's operational base. The brothers did well in the following years, and the company staff increased significantly as friends and other family members joined, including Rafael's brothers Enrique and Néstor. Yet, it was the ambitious and visionary Rafael who conceived the ingenious idea that would come to symbolize the major achievement and commercial success of the Company. This idea, which basically contemplated the development of steamship navigation along the Putumayo River, was the one that led him to undertake his epic journey from Pasto to Rio de Janeiro. The logic was simple: the Putumayo River – navigable for most of its course – would give the Company access to the Atlantic through the Amazon, thus avoiding the Pacific route, not only longer – especially to Europe – but considerably more difficult, as the cinchona bark had to be transported several days by land through the steep and craggy trails of the *cordillera*.

Reyes' plan would be completed by the construction of a bridle road from Mocoa – not far from the embarkation point at the site of Guineo – to Pasto, hence establishing a transoceanic route linking the Atlantic and Pacific oceans. Therefore, in a letter published in Pasto after he arrived from his trip to Brazil, he announced the success of his journey and also urged the convenience of the road. The French traveller Edouard André, who arrived in Pasto soon after the letter was published, mentions that the *pastusos*, following Reyes advice, asked the federal government for funds to build the road. However, notes André, 'revolution broke out and the clap of thunder vanished the illusion' (André 1984, p.773).[19]

Despite the failure of this early initiative to build the Pasto-Mocoa road, Reyes achieved a major goal during his journey, as he got permission from the Brazilian government to ship both Colombian and Brazilian goods using the Putumayo and Amazon rivers. Although Reyes would celebrate this navigation agreement with Brazil as a great nationalistic triumph and praised himself for 'having discovered an important waterway for our country' (Reyes 1986, p.161), he would later be accused of blatant self-interest. Purportedly, the Brazilian government permission was given exclusively to the Reyes brothers' Company, and stipulated that shipping was to be only for Brazilian crafts (Salamanca 1994, p.375). Although we have no record of Reyes ever having replied to such claims, it is more than likely that he would have refuted it by vehemently asserting – as he usually did – the patriotic and nationalistic character of all his individual and commercial achievements. This, on the other hand, was in perfect accordance with Reyes' character, who deemed himself a great 'civilizer' and crusader for progress and relentlessly claimed, amongst other things, to have put an end to the Brazilian indigenous slave trade in the Putumayo; 'civilizing' the 'cannibal Indians' he made contact with; and being the first Colombian to exert national sovereignty along the borders with Peru and Brazil.[20]

The significance of Reyes' self-proclaimed achievements lies, however, not so much in how patriotic or even true they were, as in how they revealed the rhetoric through which state and frontier became two dichotomous and yet mutually constructed spaces, the former's aura of authority and supremacy built upon the savagery and barbarism of the latter. It was through this very rhetoric, as previously argued, that the frontier would be assimilated to the imaginary order of the nation. Although it has already been described how this 'imaginary order' was crafted and acquired a hegemonic character, there is hardly a better graphic illustration of this order than the map of the Reyes' brothers explorations in South America (see Figure 1.4).

This map was elaborated by Reyes on the occasion of the Pan-American Conference, and published together with his paper. The civilization/savagery antinomy is here skilfully portrayed through the various features represented on the map, which together comprise a bifurcated landscape drawn on a blank sketch map of the Continent. On the one hand, we see the cross-dotted lines signalling the various explorations carried out by the Reyes brothers during the 1870s, explorations that were for the most part confined to fluvial navigation of some of the Amazon's tributaries. These lines, which mostly serve the purpose of demonstrating the navigability of these rivers, are connected to a series of square-dotted lines, indicating the projected roads – such as the Pasto-Mocoa road – connecting his colossal navigation scheme with the planned Inter-Continental railroad (indicated by the thicker bold line running north-south). This infrastructural landscape symbolizes *the future* as conceived by Reyes, a future supported by the vastness and richness of the Amazon region, and of which Reyes declared at the Conference – picking up President Roosevelt words – '[comprises] a new world that offers itself for the progress and well-being of humanity' (Reyes 1979, p.36). On the other hand, we have backward or untamed landscape symbolized by the spaces along the railways across the Andes and those in between the Amazon tributaries. The former, filled with small rings, indicates areas rich in mineral resources such as gold, silver, copper, iron and coal; the latter, a *chaotic* collage crowded with shrubs, arrows and skulls, represents the simultaneous presence of wild cocoa and rubber, 'savages' and 'cannibals', respectively. Against this composite image embodying the past, present and future of the frontier, the Reyes brothers (not shown) stand proudly at the top of the map, personifying the white man's burden of civilization and progress.

The power of Reyes' map, a power whose immanent logic cannot be detached from the colonial production of knowledge, rests not so much on the 'reality' it exposes, but on the effect through which the cartographer's *fiction* acquires an *illusion of reality*. And so we are told in the prologue to Reyes' presentation at the Mexico Conference regarding the 'accuracy' of his map:

There exist maps in abundance containing facts which have appeared in books and articles and *which are more or less real*, but how often it happened that *much is due to imagination*, such as rivers, mountains, valleys *which do not exist in reality* ... Everything contained in this valuable work has been verified by the explorer himself. Should any traveller be detained on those burning sands where the brothers Reyes dug their hollow beds under the ground find himself misled by some freak of reflection, he may rely upon the map of the Columbian [sic] traveller ... and, like Le Verrier when investigating his planet in the mysterious expanse of space [sic], may say 'I do not see it, but affirm that it exists there'. A similar *effect* is produced by this excellent map which is the result of the geographical labours of our respected and dear countryman (Reyes 1979, p.9).

But how is this illusion of reality accomplished? The 'secret' is easily revealed if we look closely at the map. There, and beneath Reyes' collage of cannibals, railroads and rubber, there is not so much chaos but

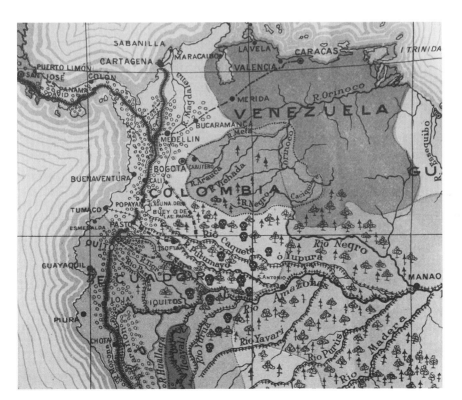

Figure 1.4 'Map showing the explorations made by the Reyes' brothers in South America and the projected Intercontinental Railroad' (detail). Source: Reyes 1902.

a clearly delineated spatial and temporal order through which the untamed space of the frontier is rendered legible by a series of simple and yet visually effective binary conventions. It is in the production of this order that the mastery of the cartographer is fully exposed, an order that, as argued by Brian Harley, can only be consummated by the multiple cartographic 'silences' – cultural, toponymic, historical – through which the 'objects outside the surveyor's classification of "reality" are excluded' (Harley 2001, p.98).

I want to emphasize the symbolic violence embedded in such silences – so profuse in Reyes' map – for it is through the enactment of this violence at the level of representation that the *physical* violence we encounter at the level of practice is both assimilated and concealed. This circular logic of enactment and concealment is what Taussig – in deciphering the rationale behind the regime of terror that unfolded around the Putumayo rubber economy – denominates the mimesis between the violence projected onto the Indians and the violence perpetrated by the *colono*: a mimesis that 'occurs by a colonial mirroring of otherness that reflects back onto the colonists the barbarity of their own social relations, but as imputed to the savagery they yearn to colonize' (Taussig 1991, p.134).

There are several episodes in Reyes' *Memoirs* where this mimetic violence comes to light. However, I will just refer to two of them, as they mirror best the indissoluble relationship between the symbolic and physical violence through which the Amazon frontier was integrated to the order of the state. The first episode took place in 1874 during Reyes' first expedition to the Putumayo, an excursion he undertook with the primary aim of spotting cinchona tree forests in the Andean foothills. After leaving Pasto and crossing the *páramo* of Bordoncillo, which separates the city from the Sibundoy Valley, he spent a few days among the Sibundoy Indians – which he described as 'semi-savages' – procuring carriers to take him across the steep trail from there to Mocoa. According to Reyes, the Sibundoy 'supreme chief', Pedro Chindoy, promised to get him the carriers in a period of five days. However, the time passed and Chindoy not only had not kept his word but had asked him to wait for another four days with the excuse that there was a party taking place in town. Although Reyes refused to accept the chief's petition, in the end he had no choice but to wait.

Finally, when the agreed day arrived, the desperate young explorer went to look for Chindoy, whom he found surrounded by 80 Indians. Then, to his surprise, the Indian chief announced to him that he had 'no real intention' of giving him carriers, and that he had better head back to Pasto before he would be forced to do so. At this point, and foreseeing the potential threat faced by him and his two 'white' companions, Reyes declared 'I realised that if I didn't make myself respected by this Indian my expedition was lost' (Reyes 1986, p.112). Thus, he narrated how he was 'forced' to pull out his revolver and make a warning shot which left the Indians 'terrified', and taking advantage of the situation he 'knocked' the chief down and with the help of his friends put him in the stocks they found in the room. Reyes' 'manoeuvre' took immediate effect as the 'frightened' Chindoy, begging him 'not to kill him', offered him ten of his best men for the following day and meanwhile entertained him with his 'best delicacies': eggs, chicken and pork meat. The triumphant Reyes, however, hardly able to get over the surprise of having frightened more than 80 Indians with only his revolver, explains the incident as due to the 'cowardice' and 'pusillanimity' of the Sibundoy Indians.

The second episode took place several months later when the brothers were already exporting the cinchona bark to Europe and the United States via the Putumayo and Amazon Rivers. During the first steam navigation in the Putumayo, Reyes mentioned that he visited and befriended the Cosacunty Indians, a tribe he found in the middle course of the Putumayo and which according to him was made up of 'around 500 beautiful and robust individuals' (Reyes 1986, p.115). Reyes felt a special affection for this indigenous community, and after spending a few days among them and having bartered some tools and chickens in exchange for wood for his steamer, he left for Belém with the promise of stopping over on his way back. 'I saw them disappear from one of the river bends' – tells the explorer in a nostalgic tone – 'and I lost myself in the immense solitude of those forests with the hope of seeing them again within the agreed time' (Reyes 1986, p.116).

Three months later and faithful to his word, Reyes arrived at the foot of the small hill where the Cosacunty lived. After having sounded the boat's whistle several times with no response from the Indians, he finally decided to climb the hill and take a look for himself. Then, as he narrates:

When we were at about one hundred meters distance from the Indians' huts, I felt an unbearable smell of putrefaction and sensed something horrible had happened to that tribe ... When I had reached the top of the hill, the smell was so nauseous that I couldn't even breath. No signs of life were seen from the huts. Accompanied by the two sailors, we rush to the chief Otuchaba's hut, whose bamboo door was ajar. I pushed it and the scene I had in front of my eyes was so horrifying that even today, after so many years, just the act of describing it terrifies me. Lying of the ground there were more than thirty corpses of elderly people, men, women and children, in a total state of decomposition. Some of them kept their eyes, throwing flames of pain and suffering (Reyes 1986, p.116).

In the midst of this scene, which Reyes himself describes as 'Dantesque' and to his stupefaction realized that it was reproduced in every hut of the village, he found the only survivors: a dying woman with a baby on her breast, who told Reyes the cause of the tragedy was an epidemic that spread among the whole tribe soon after he left. Reyes, meanwhile, concluded that the epidemic was a 'sort of tuberculosis which I have noticed the white man brings to the savages of the Amazon' and that 'this is the way these savages suffer miseries and die' (Reyes 1986, p.116).

Reyes' statement seems ironic not only as he repeatedly claimed to be the first 'white man' the 'savages' and 'cannibals' of the Putumayo had ever seen, but since it is quite likely that it was he and his crew who had spread the virus among the Cosacunty Indians. And yet, there is nothing really ironic about it. What this episode – together with that of the Sibundoy Indians – exposed, instead, is precisely the effect through which the violence rooted in the assimilation of the frontier to the state is naturalized by the myth opposing the former's state of nature to the latter's civilizing mission.

The statesman

As with every boom period, the heyday of the quinine trade and with it the glory days of the Reyes brothers' Company eventually came to an end. By 1884, and as a consequence of the growing production in Dutch plantations in Java and Ceylon, the international prices fell to the point that the brothers had to liquidate the Company due to bankruptcy. However, Reyes' final days in the Putumayo were marked by another venture that ended as a calamitous failure. Around 1880, when the quinine prices were already declining, the brothers decided to set up in the rubber business. They brought hundreds of labourers from different parts of the country and established a station in the middle course of the Putumayo River. Yet, they had barely initiated the works when an epidemic of yellow fever spread among the workers, killing in a matter of weeks about three-quarters of them. Reyes recalls this episode as a ruthless battle with a 'savage nature that defended against man's domination', and again had to make use of his revolver to persuade the frightened survivors desperately seeking to desert. This time, however, his gun proved useless against nature and the battle was eventually lost, the brothers having to abandon the place. It was also during this period that his brother Néstor, who left the station to explore other potential places for rubber, perished among the 'cannibal Huitotos'. Enrique, his other brother, died a few years later from yellow fever while extracting rubber in the Yuruá and Yavarí Rivers.

Reyes' closing words to this episode cannot hide his despair: 'from the rubber discovered by us we got nothing but disgrace and capital losses; this is the fate of the conquerors' (Reyes 1986, p.177).

* * *

The story of how the defeated entrepreneur, who left the Putumayo in 1884, became president two decades later, is long and intricate. This story, however, is to a large extent related to a successful military career he initiated soon after leaving behind his role as an explorer and businessman. Reyes' military victories, especially during the civil wars of 1885 and 1895, which he fought on the side of the Conservatives, gained him enormous popularity and also made him a prominent figure within this party.[21] Historians, moreover, tend to stress that his non-partisan character represented a major influence for his election in July 1904, particularly since the country had just emerged from the War of a Thousand Days, the most extended and devastating civil conflict since independence. Reyes' government, which he ended up exercising in an authoritative manner under the flag – borrowed from his much-admired Porfirio Díaz – of 'less politics and more administration', is generally regarded as modernist and reformist, and Reyes himself as a *man of practice* rather than a politician.[22] Among the reforms he pursued – which included the professionalization of the military, territorial re-organization aimed at counteracting the regional elites' power, and fiscal restructuring to enhance the government revenues – the modernization of the banking system and the improvement of the country's transport network occupied a relevant place. These two reforms, which were particularly directed towards creating an environment favourable to foreign investment, plainly mirror the vision the General so vigorously cultivated during his early days in the Putumayo: a vision grounded in the consummated faith in progress and in the civilizing power of infrastructure. And so he proclaimed, in a much-quoted passage from one of his speeches that:

> In times past it was the Cross or the Koran, the sword or book that accomplished the conquests of civilization; today it is the powerful locomotive, flying over the shining rail, breathing like a volcano, that awakens people to progress, well-being and liberty ... and to those who do not conform to that process it crushes beneath its wheels (cited in Bergquist 1986, p.221).

Reyes government's intentions and policies towards the frontier territories, as stated in his inaugural presidential speech delivered at the National Congress on August 7th 1904, fully embodied this vision. In a passage that implicitly refers to his own patriotic enterprises during the 1870s, he made the following allusion to the Amazon region:

> *Our* eastern territory, whose inconceivable wealth has been unveiled by a few sons of Colombia who have ventured themselves into those primary forests, or even paid with their own blood our sovereignty in those vast regions, awaits for the *efficacy* of the country's patriotism, so that, through the determined will of the entire Nation, the treasures which are currently exploited there by foreigners, in detriment of *our rights*, are opened for the country (Sánchez 1908, p.vii, emphasis added).

And so Reyes began his government by taking practical measures regarding this vast and neglected territory. In January 1905, he established the Intendancy of Putumayo with Mocoa as its capital, and during that same year the central government covered 75% of its expenses (Stanfield 1998). He actively supported the recently established Capuchin Mission, subsidizing its activities through the Ministry of Public Instruction. The Pasto-Mocoa road, a project he could not accomplish during his days as a businessman, also formed part of his concerns. In 1906, and under the justification that the road was of great importance not only in terms of economic development but of national sovereignty, he authorized, through the newly created Ministry of Public Works, engineering studies with the clear purpose of exploring and projecting potential routes.[23]

This early impetus, however, soon vanished. In March 1906, just one year after it was created, the Intendancy of Putumayo was suppressed and its territory left under the jurisdiction of the Department of Nariño. The road works were suspended by the government in early 1908, allegedly due to a lack of funds, and they would not be resumed until late 1909, by which time Reyes had left the presidency.[24] Ultimately, and in what seems a paradox, the role played in the region by the indefatigable General during the nearly five years that he remained in power was to be remembered, at best, as controversial and ambiguous. The grounds on which this perception is founded are to a large extent related to two episodes in which Reyes was directly involved.

This first episode is related to an extensive land concession granted by Reyes' government to a Colombian company for rubber extraction in the middle Putumayo, which ended up favouring the interests of the infamous Peruvian Rubber Company, the *Casa Arana*.[25] The history of the concession dates back to 1900, when Reyes was abroad as Plenipotentiary Minister to France. During that year, his nephew Florentino Calderón, who had worked in the Reyes brothers' Company in the 1870s, made an initial attempt to obtain the necessary concession from the government. The contract, drawn up by Florentino and presented to his brother Carlos Calderón, then Minister of Finance, stipulated, among other things, the cession for a period of 30 years of a vast strip of land between the Putumayo and Caquetá Rivers. In exchange, the contractor committed to support the Catholic Missions, to establish steamship navigation in the Caquetá and Putumayo and maintain the trail from Pasto to Mocoa, and to 'facilitate' the assimilation of the indigenous communities to 'civilized life'. In order to provide a legal basis for the contract, the Minister hastened to establish a Decree which specified that the central government could lease, for a period of up to 30 years, 'vacant lands' of extensions greater than 5,000 hectares. The main argument supporting the law, summarized in its first article, stated that 'the deserted regions of the Republic, home to the non-civilized indigenous population, have remained to date unproductive to the Nation' (Decree 645, 9th February 1900, reproduced in Cajiao 1900).

Lastly, and in order to avoid charges of nepotism, the Calderón brothers had asked a third person, Leopoldo Cajiao, to appear in the contract as concessionaire. However, to their surprise, at the last minute Cajiao refused to sign the contract alleging that the concession was highly detrimental to the 'interests of the Nation'. Cajiao not only accused the Calderóns of taking advantage of the country's current 'state of exception' – which had been decreed due to the ongoing civil war – to evade the legal requisite of submitting the contract to a public tender; he also argued that, once obtained, the brothers had the intention of transferring the concession to a French firm using as an intermediary a Colombian trading house based in Paris.

The scandal that followed Cajiao's charges eventually frustrated the Calderóns' initiative. But Florentino did not give up. In a brief book he published in 1902 entitled *Nuestros desiertos del Caquetá y Amazonas* [*Our Deserts of Caquetá and Amazonas*], he denounced Cajiao's accusations as an 'extraordinary defamation' and defended the concession as a 'truly patriotic' enterprise not only intended to 'civilize' the 'savages' of the Putumayo, but to exercise territorial sovereignty over the territories in dispute with Peru (Calderón 1902). With these arguments, rubber prices soaring, and his uncle now elected president, he made a second attempt. Through the firm Cano, Cuello & Co., a rubber company that was established in the Putumayo in 1903, he finally obtained from Reyes' government the much-coveted concession. The Cano, Cuello & Co. concession, signed in January 1905, granted the company ownership, for a period of 25 years, of a large territory – calculated to be 100,000 square kilometres – between the Putumayo and Caquetá rivers (Salamanca 1994, p.113). The conditions of the contract were to a large extent similar to the previous one, and included the grant of property rights at the expiration date of concession over the lands where the Company had buildings and plantations.

The Cano, Cuello & Co. concession turned into a national scandal. The scandal revolved around the fact that the concession territory, which the Company initially intended to cede to a US conglomerate – the *Amazon Colombian Rubber & Trading Company* – with the needed capital to exploit it, ended up in the hands of the Casa Arana. The scandal would have probably been kept quiet if it were not for the fact that this same territory formed a part of a dispute with Peru, a situation which, together with the recent loss of Panama, awakened nationalistic sentiments. This time the charges, moreover, directly involved Reyes, and came mostly from two of his main political detractors. Santiago Rozo, Consul at Manaus, issued in April 1910 a formal complaint to the country's General Attorney, in which he accused Fidel Cuello, Enrique Cortés and Rafael Reyes of 'Traición a la Patria' [betrayal of the nation]. Specifically, he accused the first two, general manager of Cano, Cuello & Co. and Minister to the United States during Reyes' government respectively, for having illegally negotiated the concession with the Casa Arana, and thus facilitating the *de facto* appropriation of the disputed territory. As for Reyes, Rozo denounced him as accomplice, particularly for having authorized the signature of the concession 'knowing beforehand' its detrimental effects for the country.[26]

Demetrio Salamanca, one of Reyes' former employees in the Putumayo and later Consul at Belém, went further than Rozo. He argued that the two *modus vivendi* signed between Colombia and Peru in 1905 and 1906, which recognized the *status quo* of the territory in dispute and served as a temporary measure while both countries reached a definitive agreement, favoured the violent eviction of Colombian *caucheros* by employees of the Casa Arana. As the 1906 agreement contemplated the immediate withdrawal of civil and military authorities along the zone in dispute, Salamanca explicitly suggested that this measure would also have facilitated the sale of the concession to Arana. He even went on to argue that the press censorship established by Reyes in 1906 was closely related to the Putumayo concession, thus declaring that:

> The dictator Reyes, accomplice and accessory of the crimes perpetuated in the Putumayo … had a wicked interest in keeping Colombians ignorant of the current situation in the Putumayo, and for that reason silenced the press and persecuted those patriots who defended the territorial integrity and the honour of Colombia (Salamanca 1994, p.119).

Both Rozo and Salamanca's claims contain numerous rumours and versions taken from other people involved in the concession, referring to juicy bribes, blackmail and fraud. Through them, the case itself becomes an intricate game of mutual accusations when the perpetrators themselves turn into victims and victims into perpetrators, thus blurring the line between fiction and fact. Reyes himself, who never set foot in the Putumayo after his tragic rubber venture, refuted the criticisms and accusations by invoking once more the patriotic crusades he accomplished during his youth, and the many sacrifices he and his brothers endured among the 'savages' and unhealthy forests (Reyes 1912, 1913). Eventually, the concession scandal subsided and Reyes, along with the other accused, escaped unpunished, an outcome typical of the country's countless episodes of political corruption. After having resigned the presidency in June 1909 – due mostly to the increasing opposition to his dictatorial regime and unpopular measures regarding the negotiations with the United States in relation to Panama – he sailed into exile in Europe on board a United Fruit Company boat. Whilst there, and just as another scandal of much bigger proportions unfolded around the atrocities involved in the midst of the rubber boom, the General devoted his time to spreading word of his heroic adventures and promising discoveries in the 'New World' of the Putumayo.

* * *

The relevance of the Cano, Cuello & Co. concession scandal to the argument put forward here lies not in how fabricated or real the claims were or to what extent justice was done, but in how these claims (revolving around charges such as betrayal of the 'honour of the nation', the lack of 'patriotic virtue' or simply bribery and corruption) silenced or concealed the violence through which the frontier was constituted. The apparently ambiguous and controversial Reyes government's attitude towards the Putumayo must be appraised precisely in that sense: not as resulting from inconsistent behaviour on the part of the ambitious and egocentric president, but as reflective of the discursive and material practices through which this territory was incorporated to the order of the state. Let me finish by briefly referring to the other episode for which Reyes' government would be remembered and which elucidates this point.

In early 1906, after being victim of a failed *coup d'état* and attempted murder, Reyes decided to exile the main conspirators in the Putumayo. For this purpose, he established by executive decree the Penal colony of Mocoa in April of that year, so the town where he and his brothers initiated their 'patriotic' quinine business back in 1870s, ended up serving as a site of confinement for his political adversaries. Although most of the conspirators would soon be released, Reyes' decision would leave an indelible imprint on the region's history. Guillermo Guerrero, an engineer from Mocoa who deeply admires Reyes' visionary interoceanic project and argues that the presence of the state in the Amazon began with him, could not hide his bewilderment and indignation about his decision to make Mocoa a penal colony. This is the angry reply he gave me the first time I asked him about this event:

> I haven't been able to read that decree just because it gets on my nerves, although this is my own emotional problem … Execrating Mocoa, demonising Mocoa as a penal colony. So one thinks, how dumb the people from the Colombian Andes. They never had a vision of the Amazon. They always saw the Amazon, let's say, as shit. Where is the Amazon? It's *in* the shit.

Guillermo's resentment must surely be shared among many people inhabiting frontier regions across Colombia. For, although Reyes can be considered a pioneer in conceiving these territories as 'natural' places to exile criminals and political prisoners – he also established other penal colonies in the Llanos region (Rausch 1999, pp.302–304) – this constituted a common state policy throughout the twentieth century. Moreover, and despite the fact that many of the initiatives of creating penal colonies during this period remained on paper, it has been noted that they contributed to create an image of regions such as the Llanos or the Amazon as 'space[s] of exile' (Gómez 2011). Still, and against Guillermo's perception, one could say that the people from the Andes did actually have a *vision*. This vision, whose origins date back to colonial times and which we find perfected in the nineteenth century by *criollo* figures such as Caldas, Codazzi and Reyes, is essential in understanding the myth on which the modern state was founded and sustained.

As I argued, the apparently contradictory nature of this myth lies in that, although state and frontier *appear* as two fundamentally irreconcilable orders, the latter's state of nature becomes not an *obstacle* but a condition of possibility in which the former's power is rooted and perpetuated. Thus, and underneath the illusory effect by which the frontier appears in a relationship of externality to the order of the state, we find not its isolation or exclusion from that order but the logics and practices through which it is included in it.

It is only by considering this relationship of inclusive exclusion in which state and frontier have been historically entangled, that I argue it is possible to understand many of the plans and projects, utopian or not, that have shaped or transformed the Amazon's physical and social landscape. It is also from this background, as I seek to show in the next two chapters, that we can come to see the story of the road not as an isolated event but as a reflection of the spatial and historical process of state and frontier-making in this region.

Notes

1   The specific expedition Reyes referred to in the talk was the long journey he made by himself from the city of Pasto (Colombia) to Rio de Janeiro during the years of 1874 and 1875 while he was searching for an export route for quinine and rubber. However, his presentation at the Pan-American Conference is largely a composite account of the numerous trips he made in the Putumayo in the company of his brothers between 1870 and 1884. This is indicated by the fact that the narrative is presented in third person (suggesting the presence of his brothers), and also since it incorporates events that took place on other expeditions. A complete account of the years Reyes spent in the Putumayo is contained in a series of letters he wrote to his sons during the early 1910s, published posthumously as his Memoirs (see Reyes 1986).

2   Reyes was a contemporary of Stanley, a figure whom he admired together with David Livingstone. He actually made reference to them in his chronicle (Reyes 1979, p.14), comparing the Welsh-American and Scottish explorers' expeditions in Africa to his South American version. This parallel is worth mentioning since the figure and writings of Reyes can be better understood in the context of the relationship between nineteenth-century geographical exploration and imperialism. For a reading of Reyes in this particular respect see Martínez (2013). For a broader discussion of this subject in the context of European imperialism see, amongst others: Bell, Butlin and Heffernan (1995); Driver (2001); and Godlewska and Smith (1994).

3   The *encomienda* was, in broad terms, a system of tribute extraction under the form of labour established by the Spanish crown in America. In the case of Mocoa, the establishment of *encomiendas* was directly associated with the development of gold mining in its surroundings throughout the sixteenth and seventeenth centuries (Ramírez 1996, pp.80–81).

4   This situation was to a large extent shared among the Spanish territories across the Andes from Colombia to Bolivia (see Belaúnde 1994).

5   For an account of the different foundations of Mocoa throughout the colonial period see Mora (1997, pp. 44–46).

6   The history of pre-Hispanic, colonial and early republican exchange routes between the Colombian Andes and the Amazon constitutes a relatively well documented topic. See, for example, Gómez (1996); Gómez and Domínguez (1995); Ramírez (1996); Ramírez and Alzate (1995); and Uribe (1986, 1995).

7   As noted by Ramírez (1996, p.109), although the overload of human carriers was condemned by law as early as 1542, the custom of travelling on the back of the indigenous peoples continued to be a common practice until well into the twentieth century. For a discussion on the practice of *silleros* in the broader context of colonialism see Taussig (1991, pp.287–335).

8   The area of the Colombian Amazon, which amounts to 413.473 km² distributed between seven departments, corresponds roughly to one-third of the country's total area.

9   Although the *Uti possidetis iure* principle would be the subject of subsequent disputes, its adoption following the end of the Spanish rule had the main purpose of preventing fratricidal struggles amongst the independent states.

10   'Informe del cónsul de Belém enviado al Ministro de Relaciones Exteriores'. Belém del Pará, 12th June 1894. AGN, Archivo diplomático y consular-MRE, Box 127, Folder 277, fols.7–8.

11   The term *creole* (from the Spanish *criollo*), alludes to the name which was originally given to all the descendants of Spaniards born in America.

12   On the issue of race, state-building and nationalism in Latin America see: Larson (2004); Stepan (1991); and Wade (1997). Some bibliographical references on this subject for Colombia include Arias (2005); König (1984); Safford (1991); and Wade (1993).

13   The role of geographical discourse in the creole elite's project of nation state has been discussed by Arias (2005), Múnera (2005), and Nieto (2008), among others.

14   A soon as 1823, a state law demanded local governors to support the settlement of foreigners in the 'most advantageous lands'. Years later, in 1847, the government sanctioned an Immigration Law which contemplated extensive benefits, and stated that the executive power could dispose of around two million hectares of 'unoccupied lands' with the purpose of granting them to foreign immigrants. For a general description of attitudes and government efforts towards foreign immigration in Colombia throughout the nineteenth century see Martínez (1997).

15   These sources correspond to the accounts elaborated by José María Quintero (Prefect of Mocoa), the Presbyter Manuel María Albis and Pedro Mosquera, *Corregidor* (mayor) of Mesaya.

16   Throughout the nineteenth century Colombia went through several names and territorial restructuring. From 1830 to 1856, it adopted the name of Nueva Granada.

17   For a historical account of the quinine boom and the role played by the Reyes brothers' Company in this context, see: Domínguez (2005); Gómez and Domínguez (1995); and Zárate (2001).

18   The Putumayo rubber boom constitutes a subject that has been extensively studied. Some relevant works on this subject are: Domínguez (2005, pp.79–200); Gómez, Lesmes and Rocha (1995); Pineda (1987); and Taussig (1991); Stanfield (1998).

19   André is alluding here to the 1876 civil war, one among several conflicts that confronted Liberals and Conservatives throughout the nineteenth century.

20   These claims are found in numerous occasions in Reyes' writings (see, for example, Reyes,'Carta al Ministro de Colombia en Washington', Bogotá, julio 25 de 1905, BLAA, Libros raros y manuscritos, MSS391; Reyes 1986, p.135, 142–143; Reyes 1920, pp. 580–589.

21 A detailed account of Reyes' military career can be found in his biography by Eduardo Lemaitre (1981, pp. 15–52, 122–163).
22 For a discussion and description of Reyes' government policies and reforms see: Bergquist (1986); Bushnell (1993, pp. 151–161); Mesa (1986); Sánchez (1908); and Vélez (1989).
23 AGN, Ministerio de Obras Públicas (Ministry of Public Works, henceforth MOP), vol. 1407, fols.8–10, 39–40.
24 AGN, MOP, vol. 1407, fol.154.
25 See Gómez (1993) and Salamanca (1994) for a detailed account of this concession.
26 The complete text of Rozo's formal complaint is reproduced in Gómez (1993, pp.21–23).

2

A Titans' work

*Allí es donde mejor pueden apreciarse el poder del hombre y la fuerza de
la dinamita* [There is where the power of man and the force of dynamite
can be best appreciated]

(Gutiérrez 1921, p340).

A mission's tale

During the first three decades of the twentieth century Bogotá witnessed
a rapid growth. Its population, which barely amounted to 40,000 in
1870, increased almost sixfold during the following 50 years, reaching
235,702 in the 1928 national census, a change that largely mirrors an
emerging trend towards nation-wide urbanization. The city's rural
fringes began to be transformed into paved avenues and residential
neighbourhoods, with the result that the urbanized area more than tripled during this period. This demographic and spatial expansion, which
is often considered to mark the beginning of the transition of Bogotá
from a 'village' to a 'modern metropolis', also consolidated the city as
the country's financial, political and demographic centre (Misión
Colombia 1988).

Bogotá's fast pace of urban growth, however, contrasted strikingly
with its geographical isolation from the rest of Colombia, one of its most
distinctive features from the time the city was founded in 1538. Although
it was finally connected in 1909 by rail to the Magdalena River via the
Girardot railroad – a 132-kilometre line through the rugged Andean

Frontier Road: Power, History, and the Everyday State in the Colombian Amazon,
First Edition. Simón Uribe.
© 2017 John Wiley & Sons Ltd. Published 2017 by John Wiley & Sons Ltd.

topography whose construction took nearly 30 years – its communication with most of the country's main cities continued to depend on the nineteenth century's precarious and torturous trail network. Even in 1934, when the national government had begun to alter its transport policy from railways to road and highway construction, the country's railway network – largely comprised of short and unconnected lines – barely exceeded 3,000 kilometres, the second shortest railway length per capita of Latin America (Pachón and Ramírez 2006, pp.28–29). In addition, the first roads, which began to be built during the 1900s, consisted mostly of narrow dirt routes that followed the paths of the old republican trails, themselves inherited from colonial times, and lacked any technical design.

Against this backdrop, the observation made by Donald Barnhart, a US scholar who studied the country's transport modernizing policies during the 1930 and 1940s, comes as little surprise. Whilst noting that the Bogotá of the early 1920s was one of the most isolated capitals in the world, Barnhart observed that the 'towering mountain ranges which enveloped the people limited their outlook and restricted their movement, while the plodding pace of the mule seemed to set the tempo of their life' (Barnhart 1985, p.1). Nevertheless, this peculiar form of myopia, particularly acute amongst the *criollo* elites from Bogotá, did not solely stem from the fact that they were enclosed by 'towering mountain ranges' blocking their view towards the neighbouring lowlands, not to mention the peripheral territories of the country. More telling is the name won by the city in the late-nineteenth century, the 'Athens of South America', referring to the literary erudition of its elites, who imagined themselves more inclined to write and engage in conversations about poetry and philosophy than about the harsh realities of the country.

In this 'milieu', it might be expected that news brought from remote country regions would have an alluring effect on the cosmopolitan Bogotanos, far more familiar with Paris, London or New York than with the Guajira, the Llanos or the Amazon. Thus, that an event such as The National Congress of Missions celebrated in the capital in August 1924 did not pass unnoticed is not surprising. The Congress was preceded by a massive exhibition of indigenous objects gathered by the missionaries, as well as all kinds of objects – clothes, toys, ornaments – collected as donations for the Indians and missionaries. The two sections, according to a chronicle of the event, were remarkably tied by a linear timeline, the former incarnating the origins of 'barbarism' and the latter the triumph of 'civilization' (Boletín 1924, p.264). However, the section that aroused more interest, and the one that made the exhibition a 'noisy success', was the indigenous one. There, the public could not only appreciate the 'primitive, rudimentary, backward and miserable state' of the indigenous tools, weapons and other objects, but marvel at the 'products of a rich and exuberant nature' and the 'work done by the missionaries' among the indigenous children, 'the hope of a promising and not far future' (Boletín 1924, p.261).

This preamble was followed by the Congress itself, one of the main aims of which was to offer public lectures about the challenges faced and the progress made by the Catholic missionaries scattered across the 'Mission territories' (regions under jurisdiction of the Church), which at the time encompassed about three-quarters of the country. The most popular lectures consisted of 'night sessions' in the Faenza Theatre (an *art nouveau* building inaugurated months before), where the lecturers entertained the public –composed mostly of distinguished ladies and gentlemen from the capital – with tales and photographic projections from the 'savage' lands of Colombia. The enthusiastic chronicler, referring to the great success of these sessions, fervently stated that:

> Never had the Republic been known so well than in those memorable nights; those conferences were living lessons of geography, religion, national history; we can say that we visited personally those vast regions which form three quarters of the Republic; our eyes saw and our ears heard what exists and happens there (Boletín 1924, p.265).

Among the speakers was the Catalan Capuchin friar Fidel de Montclar (see Figure 2.1), Apostolic Prefect of the Putumayo and Caquetá.[1] Known for his energetic character and stern charisma, the 56-year-old, white-bearded Friar had been head of the Mission for 20 years, almost as long as the Capuchins had been established in the region. His speech, suggestively entitled 'What the Putumayo and Caquetá territories were before the creation of the Apostolic Prefecture; what they are nowadays, and what they can become', was essentially a semi-fictional account of the 'radical transformation' of the region brought about by the Catalan Mission (Montclar 1924). Skilfully crafted, the story had as central characters two imaginary friars, a narrative artifice that allowed Montclar to dramatize the history of the Mission in the form of a short tale. And the beginning of this tale, not unexpectedly, was quite similar to Reyes' opening of his epic adventure in the Putumayo given years back in Mexico – except in this case the tale's hero is not the megalomaniac explorer in chase of fortune but the humble missionary in search of souls.

Leaving behind the city of Pasto, Fray de Montclar began, two Capuchins riding on modest mounts through a craggy trail intermittently looked towards a chain of mountains that is perennially surrounded by

Figure 2.1 Fray Fidel de Montclar, n.d.
Source: Archivo de la Diócesis de Sibundoy.

dense fogs. Eventually, at the point where the road ends, they left their mounts behind and began the excruciating ascent of the ridge:

> In front of them, there was an imposing *cordillera* that must be crossed to reach the Mission. They ask: Where is the gate? Which is the road? The answer is that they must open the road and force the gate. On the other side of those high summits, where every human consolation looks distant, a mysterious region is revealed. There, sadness and melancholy may settle; in those dismal cliffs, they even believe they can see written the sinister words Dante read at the gate of another fateful place. Their missionary vocation is put to a hard test (Montclar 1924, pp.5–6).

During the many hours of walking through the freezing *páramo*, continued Fray de Montclar, their feet bleeding, stiff with cold and stunned by rain and wind, the exhausted monks 'attempt with extraordinary effort to overcome the barrier which separates two worlds: the civilized and the savage' (Montclar 1924, p.6). And then, when they finally begin to descend the opposite slope of the *cordillera*, they glimpse

Not the promised land where honey and milk flows, but an incommensurable and tangled forest where the puma and the jaguar, in partnership
with enormous and poisonous snakes, dispute the domain of those uncultured lands with the Indians (Montclar 1924, p.7).

In his despair, one of the friars sat on the trunk of a gigantic fallen tree,
and from there he strove to make sense of the vast country given to him
as mission territory. He anxiously browsed through some documents he
had brought with him that narrated the numerous failed missions there
in previous centuries. He also looked to Reyes' story, admired his tenacity
and endurance, and was struck by the courage of other *caucheros* who,
like the famed explorer, went into the forest, dominated the 'ferocious
tribes of cannibals' and extracted huge quantities of rubber. Then, leaving aside the documents, the friar wondered:

What curse lies over this fateful land for so many sacrifices, trials and
combined efforts to have failed? Nor the heroism and devotion of the
missionaries, nor the adventurous spirit of the *caucheros*, have done
anything: everything has *vanished like an illusion*, and the Caquetá and
Putumayo have again been wrapped in the shadows of a dark night
without any sign of the dawn of a beautiful day! (Montclar 1924, p.8,
emphasis added).

Nevertheless, and despite this gloomy picture, the prostrated friar
does not give up. He made projects and planned new excursions: he
traversed swamps, waded across turbulent rivers, slept under the trees,
opened trails with a machete through the dense forest. And then, after
months of deprivations and perils, he finally found an indigenous
tribe. Many of the Indians were totally naked and the friar didn't
speak their language. But he knew that in order for the 'light of faith'
to shine in those 'dark minds' he had to adapt to the hostile nature of
the jungle. He eventually did, yet not without great difficulty, and
began to harvest the seeds he planted with so much effort in that
'infertile soil'.
    A few years have passed, the Prefect tells us, and the 'sole of the mis-
sionary has trodden the jungle in every direction, and the sweat of his
brow and the tears of his eyes have fertilised those solitudes' (Montclar
1924, p.11). New missionaries came to join him, and with their joint
efforts the Mission began to flourish: new towns were founded, a few
churches and schools were built, and the gospel is spread rapidly among
the natives. However, warned Montclar, although the Mission's progress
has marked the beginning of 'a radical transformation in those jungles',
a barrier of steep mountains, deep marshes and icy *páramos* kept this

territory isolated from the civilized world. If for any reason, he wondered, the missionaries had to abandon the Putumayo, all their efforts so far would have been in vain. Thus, and evoking the Capuchin's major achievement in the Putumayo, he reached the climax of the story:

> It is necessary, then, to overcome the impossible: the mountains have to be flattened, the rocks blown, and the *cordilleras* demolished. The age of miracles has not come to an end. The missionaries are transformed into sappers and the priests into engineers, and they undertake what had been deemed impossible: the opening of a road from Pasto to the Putumayo through the Andes, over icy *páramos* and granite mountains. The miracle happened, and the mountains were flattened, the valleys filled, the cliffs blown and the *cordilleras* opened (Montclar 1924, p.13).

This passionate ode to the fused creative destructive power of Christian faith and dynamite, endorsed by celebratory quotes to the Capuchin's 'prodigious work', marks a break in the Prefect's tale. What follows is largely a description of the Capuchin's achievements 'once the mountains' fence that enclosed and isolated the Putumayo was broken': of how they founded agricultural colonies and promoted colonization throughout the region, established numerous schools and increased the number of missionaries, resuscitated Reyes' project of opening a navigation route along the Putumayo River, and so forth. Approaching the end, in a passage that strongly reminds us of Reyes' flamboyant accounts of the Amazon forest, Montclar excited the audience by portraying the most exuberant future for the region, a future sustained by its 'grandiose' and 'vacant' nature. That portrait, he further declared, as if pretending to wake up a hypnotized crowd, 'is not a fiction'. Yet, in order for this promising future to come true and the illusion not to vanish, he concluded by listing a number of requirements: first and foremost that 'Colombia does not neglect the material and moral support of the Mission' (Montclar 1924, p.23).

* * *

There are three elements in Montclar's story that I would like to highlight as they are central to the purposes of this chapter. First, there is the circular logic in which the tale itself is immersed. As in the case of Reyes, although a quarter of a century later, Montclar framed his story following a linear temporal pattern that clearly differentiates past, present and future, a pattern marked by the strenuous yet inexorable transition of the Amazon frontier from a state of 'savagery' to one of 'civilization'. However, the tale has no ending. For, as Montclar emphatically warns, without the Mission, no matter the 'progress' accomplished so far, the fate of the Putumayo faces an inescapable return to its aboriginal, 'savage' state. As we shall see later, every time the Prefect saw the Mission's *status quo* or interests threatened, he appealed to that rhetorical formula in his letters and reports to the national government. Víctor Bonilla, in a book that strongly questioned the legacy of the Capuchin Mission and unleashed a scandal of international proportions in the late 1960s, captures sharply the rationale behind the formula. In noting how Montclar often exalted the 'savage' character of indigenous communities that, by the Republic's legal standards, had long been 'civilized', he observed that, had the Prefect adopted a different posture, 'he would have had to reconsider the whole basis of the Capuchin Mission's civilizing action' (Bonilla 1972, p.176).

I would like to relocate this rationale from the particular history and legacy of the Mission to the broader discussion about the relationship between frontier and state. As discussed in the previous chapter, it is in the perpetuation of the binary opposition between 'civilization' and 'savagery' and other analogous dualisms that we find one of the essential elements within which the state's illusion of legitimacy is ultimately grounded and sustained. This is precisely the boundary-marking effect that Montclar's tale intends to achieve in the audience of the Faenza Theatre. Put differently, by warning the public of the radical yet fragile civilizing effect of Christian faith and material progress in the frontier, the tale also serves as a reminder that the boundary has to be maintained so that the civilizing mission of the state, embodied here in the Capuchin friars, can be preserved.

This leads us to a second element, namely the state-character that the Mission assumes throughout the tale. Not only does its central character appear as an abstract entity – an anonymous friar, although we know it alludes to the Capuchin Mission and more specifically to Montclar himself – but the friar seems to act as a sovereign force transforming the savage nature of the frontier. Certainly, Montclar makes no single mention of any other kind of authority in the Putumayo and Caquetá other than that exerted by the Mission. Just as in Reyes' depiction of the *Territorio del Caquetá* in the early 1870s as a *terra incognita*, the Prefect's omission might seem an overstatement. As noted previously, the civil authorities, although negligible, had been present in this territory since the mid-nineteenth century. During the first two decades of the twentieth century (the period encompassed in Montclar's tale), along with the missionaries, other state agents existed such as the *Comisario especial* (the principal civil authority), a small police force and occasional military contingents sent due to the ongoing conflict with Peru. Yet the great political, economic and spiritual power that the Capuchins had in the region during that period is hardly questionable. The political regime in force, which extended from 1886 to 1930 and was marked by the political hegemony of the Conservative Party, had granted ample authority and influence to the Catholic Church. The Concordat of 1887, signed between the Colombian government and the Vatican, officially recognized the Catholic religion as a keystone of the nation's social and political order. Furthermore, government regulations such as the Law 89 of 1890 and Law 72 of 1892, stipulated that the ordinary laws of the republic did not apply to the 'savages', who occupied roughly three-quarters of the country. Those were left under the jurisdiction of the Catholic missions, which in order to secure its 'reduction to civilized life' would be given 'extraordinary' civil, judicial and criminal authority.[2]

The laws mentioned were grounded in a politico-juridical framework dating from the 1840s, which essentially turned the country's peripheral regions into spaces of exception. The Constitution of 1843, for instance, established that places which could not be incorporated into any of the country's provinces due to their remoteness, would be constituted as *Territorios nacionales* [national territories], territorial units governed by 'special laws' issued by the executive power (Rausch 1999, p.76). This basically meant that about two-thirds of the country remained outside the constitutional law under the pretext of their geographical isolation and uncivilized state. Although the legislation and regulations regarding these territories were often modified by successive governments, their exceptional legal and political status remained in force well into the twentieth century.

This order of things conferred great power to the Capuchin Mission. While this dominant *status quo* of the Mission is hardly ever overlooked in the literature, the interpretations vary: some argue that the state 'delegated' the task of 'civilizing' indigenous peoples to the Mission or, more generally, that it was part of a state policy to integrate the frontier territories into the nation (e.g. Fajardo 1996, p.264; Gómez 2011, p.205); others, that the missionaries basically filled a 'power vacuum' left by the state, too weak to reach remote regions of the country, otherwise deemed of little political and economic interest (Brücher 1970, pp.34–35; Palacio 2006, p.119; Palacios 2007, p.74); while others go on to say that the Mission and the state represented two antagonistic powers (Casas 1999, pp.212–216), or even that the former took over the latter's monopoly of power and eventually became 'a state within a state' (Bonilla 1972, p.212). As the Mission actually appeared and acted variously as a state ally, emissary or even adversary, these views are not necessarily exclusive. However, they all conceive both entities as two separate and independent realms, or assume the state as an abstract construction existing in an external relation to the material existence of the Mission.

The view I am proposing is different. I would like to suggest that we cannot consider the spatial and historical process of state-building in the Amazon without conceiving the Mission as an inherent element of this process. This argument is basically grounded upon two ideas discussed in the previous chapter: first, that in studying the state we have to account for the practices and infrastructures through which it is configured and encountered. Secondly, that upon addressing the (inclusive-exclusive) relationship between state and frontier, we have to pay special attention to the different actors through which this relationship has been temporally and spatially constituted. Thus, and as in the case of Reyes, it is in this sense that the Capuchin missionaries are addressed here: as personifications or expressions of certain *hegemonic* discursive and physical practices through which this relationship has been constructed.

A third element I would like to highlight is the central role that the road plays in the tale. In fact, its whole plot, framed around this single event, tells how in breaking a *spatial* barrier – the Andean *cordillera* – the road simultaneously overcomes a *temporal* one – that between 'savagery' and 'civilization'. Montclar, moreover, insistently warns us that the 'miracle' of the road is not an 'illusion', and in doing so he portrays it as a ground-breaking event in the history of the Amazon. What the Prefect does not tell his audience, is how far the actual history of the road is from the tale. Yet, as in Reyes' map, the crucial aspect of Montclar's tale lies not so much in how fictitious or real it is, but in the effect by which the fiction acquires an *illusion of reality*. When looking at the history of this particular infrastructure, to which I now turn, I suggest that it is only by considering this effect that we can fully grasp the significance of this infrastructure in the spatial history of the frontier.

The General's last sigh

The tragic events that marked the end of the Reyes Brothers' Company in the Putumayo in the early 1880s did not persuade the young Rafael to give up his dream of opening a route linking the Atlantic and Pacific oceans. So, two decades later, now president, he not only resuscitated the transoceanic project; the way he embraced it – more as a personal quest than a government plan – indicates that he was determined to realize it. Well aware that the project's main obstacle had always been the steep mountains hindering the access to the Amazon lowlands, the General focused his energies on overcoming the infamous 'Andean barrier'. Thus, during his first year of government, he commissioned Miguel Triana, an engineer from Bogotá, to study the best route for a bridle road from the Andean city of Pasto to La Sofía, the small port he had founded on the banks of the Putumayo back in the 1870s.

Triana spent the first half of 1906 in the Putumayo and wrote an extensive account of his expedition in a book dedicated to Reyes, whom he honoured with the title of 'the distinguished explorer of the Caquetá' (Triana 1950). Nevertheless Triana, a liberal from the capital's elite, unlike Reyes, was not a fervent devotee of the Spanish friars, about whom he commented with scorn that 'their hearts are full of hatred against the Republic and lack of love for the country' (Triana 1950, p.86). Thus, the engineer was not persuaded by the Capuchin friars he met in Mocoa and whom, he said, insistently begged him to recommend to the government a route through this town and the Sibundoy Valley. In fact, in the final report submitted to the Minister of Public Works, Triana made a detailed analysis of the possible routes, leaving the one passing through Mocoa – a 'ruined village' in his view – as the least attractive option in terms of costs.[3]

Triana might have underestimated the anxious appeals of the missionaries. Just a few days had passed after he had presented his report when the Minister got a letter from Julián Bucheli, governor of Nariño and close friend of Montclar.[4] In the letter, Bucheli persuasively lobbied for the Sibundoy route, arguing that others, although shorter, would pass through 'entirely deserted regions'. The former, on the contrary, he added, in connecting Pasto (capital of Nariño) with the densely populated Sibundoy Valley, offered two indisputable advantages: first, it would encourage the colonization of the valley, thus putting its 'semi-savage' inhabitants in contact with the peoples of Nariño; secondly, as the Capuchins had assured him, the same Indians would be an abundant source of 'free labour' for the road.[5]

Whether Bucheli's missive influenced the Minister or there were other interests at stake is hard to say. The fact is that Triana, who according to the Decree that set the terms of his contract was expected to proceed with the layout of the road once he concluded his survey,[6] got lost from the picture. Therefore, and with Triana's report rendered useless, the project was halted. The restless president, however, reacted soon after this failed attempt. Through a series of hasty telegrams to Bucheli, he recommended new engineers, allocated resources and above all urged the construction of the road to start without delay, repeatedly stating its crucial importance both commercially and in terms of national sovereignty.[7]

In February 1907, after two other engineers had refused the job, the contract for the complete design and construction of the road was finally signed. The contract stipulated that the road should pass through the Sibundoy Valley.[8] Although it took the new contractor, Víctor Triana, a whole year to design the complete route, he only concluded a small section of the layout, basically consisting of a one-metre wide trail and some forest clearing around it.[9] However, and contradicting the recommendations made by Miguel Triana two years earlier, he chose the route from Sibundoy to La Sofía that passed through Mocoa (see Figure 2.2). Reyes, meanwhile, grumbled from Bogotá about the slow pace of the works. Without hiding his frustration, in January 1908 he wrote a telegram to Bucheli in which he declared: 'With regret, I note that too little has been done at a great price'.[10] He was probably right and yet, his authoritative voice, as well as his government, had already begun to weaken. So, when in May of that same year a Presidential agreement suddenly decreed that the road works should be suspended indefinitely due to the lack of funds,[11] it is quite likely that there was little that the General could say or do. In June 1909, his dream unfulfilled, Reyes went into exile and would not return until 1918. As for his cherished port of La Sofía, according to his main biographer, it would soon be erased from the country's official map (Lemaitre 1981, p.363).

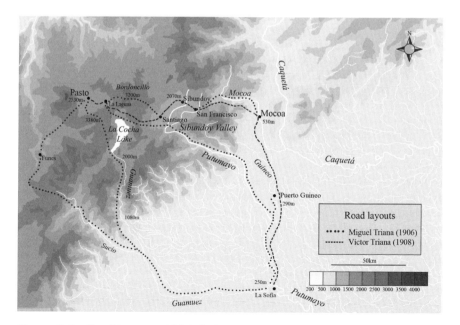

Figure 2.2 Road layouts, 1906–1908.
Source: Elaborated by author. Based on Triana (1950, pp.381–91) and AGN, MOP, Vol.1407, fol.167.

The odyssey

By mid-1909, and after three years of surveys, unfinished contracts and bureaucratic delays, the road project had not progressed much beyond the few kilometres of the abandoned narrow trail left by Víctor Triana. The discussions that led to the selection of the route, however, had exposed an underlying political economy where different interests converged. That the Capuchin route though Sibundoy and Mocoa ended up prevailing is not surprising. For, neither Reyes the businessman and explorer (obsessed with the idea that the road should reach La Sofía) nor the statesman (aware of its strategic value to exercise sovereignty over a territory in dispute with Peru) was much concerned about the actual route as long as the road was completed. Most of Nariño's economic and political elite, attracted by the lands of the Sibundoy Valley and the commercial and extractive opportunities offered by the future road, saw in the Catalan Mission a natural ally to pursue their own interests. Still, if to the former the road largely meant a long-awaited gateway to a promised land of inexhaustible resources, to the latter it represented an infrastructure vital in accomplishing their civilizing mission.

As early as 1893, when the first Capuchin expedition to the Putumayo took place, the difficult access was perceived and experienced as the biggest difficulty to the foreign catechumens. Angel Villava, who took part in the referred to expedition, kept a journal where he described in detail the hardships they suffered crossing the *páramo* of Bordoncillo and the steep trail from the Sibundoy Valley to Mocoa. 'In this rich and fertile country', he wrote, whilst noting that its major problem was the lack of roads 'there is no sign of progress: everything remains the same or even worse than one or two centuries ago' (Villava 1895, p.14). A decade later, when the Capuchins had already established the Mission's headquarters in Sibundoy and the Apostolic Prefecture of the Putumayo had been erected, the situation had barely improved. Fray Jacinto de Quito, who arrived in 1903 and remained with the Mission for more than 50 years, devoted a chapter of his memoirs to describe the travel conditions during its early times. Suggestively entitled 'On Indian's back', he observed that in order to traverse the Bordoncillo, it was 'imperative' to resort to the 'very ancient method' of riding on the backs of the Sibundoy Indians, for it was 'materially impossible' to employ beasts of burden on those trails. The brunt, he added, was to be borne not by the *sillero* but by his passenger, since the former 'walked calm and content, thinking only of the fee he would be paid, and willing to repeat the same trip for a thousandth time' (Quito 1938, pp.25–27).

As noted in the previous chapter, the Andes' topography never represented a barrier to the native inhabitants of the highlands, the foothills or the lowlands, who for a long time had maintained an active cultural and economic exchange. Their intricate system of paths and trails, moreover, acted also as a mechanism of resistance, not only facilitating the evasion of authority but also hindering the access of foreigners to their territory. The natives seem to have been very conscious of this, even by the time the Capuchins had been established there for a while. Hence, Miguel Triana, who on his way back to Pasto took the trail from Mocoa to Sibundoy, mentioned how his Indian carrier followed a trail through the left margin of the Mocoa River, much more rugged and longer than the one opened by the 'whites' on its other margin. The Indians, pointed out Triana, following the 'tradition of their ancestors', intentionally discarded the whites' trail (Triana 1950, pp.356–357). Yet Montclar, well aware that a road would do much more than enable the access of a few missionaries to the vast territory under his jurisdiction, was also very conscious of this.

In 1909, and witnessing how Reyes' project had been wrecked, the energetic Prefect wrote a long letter to the Minister of Government. The document, which as with many of his writings constitutes a fine work of rhetoric, imbued the road with a paramount significance where its many attributes appeared masterly entwined: national sovereignty, colonization, access to boundless riches, civilization and progress.[12] The road, moreover, he reassured the Minister, did not require 'major sacrifices' to the nation, as it was 'one of the easiest works to accomplish'. Thus, he mentioned how without any financial assistance, the missionaries themselves had opened – with the help of some Indians they had 'induced' to work – more than 50 kilometres of bridle path between the Sibundoy Valley and Pasto. 'What would have we accomplished with aid from the government?' he asked the Minister, while at the same time assuring them that with a contribution of 40,000 pesos the Mission would finish the road to Sibundoy and also open the section from there to Mocoa. Yet, and in order to stress the derisory character of the amount requested, the Prefect finished with an admonitory message: as chief of the Mission, main ecclesiastic authority of the Putumayo and Caquetá and 'advocate of Colombia's integrity', he asked the Minister 'to inform the Nation of the imminent danger it faces of losing those rich and beautiful territories'.

It took the Minister less than a month to get Montclar's request approved. Through Resolution 21 of September 1909, the government commissioned the Governor of Nariño to control and oversee the funds for the road, and appointed Montclar as General Director of Works. The latter, absent at that moment, promptly delegated the task to his colleague Estanislao de Las Corts, and the works began on 25th October.

The friars soon faced the problem of finding labourers. The Mission, as the Prefect had stated to the Minister in his letter, had certainly initiated the construction of a road from Pasto to Sibundoy three years previously. In November of that year, Montclar had printed a pamphlet in Pasto entitled 'The savages of Caquetá, Putumayo, and the road to Mocoa', in which he appealed to public charity to collect funds for the works.[13] According to the Prefect, since the amount gathered did not reach the 200 pesos needed, the missionaries had no choice but to undertake the project themselves, resorting to the 'tributary work' of the indigenous of the Sibundoy Valley and La Laguna, an indigenous town on the outskirts of Pasto. This form of compulsory labour, which basically consisted of the obligation of indigenous communities under the tutelage of the Mission to work for free for certain number of days a month, did not cease when the government began funding the construction of the road. That this practice actually intensified is evidenced by the fact that a few weeks after the works had been initiated, and due to the recurrent claims of the Indians, the provincial government of Nariño issued a resolution forbidding the Mission to make use of this system for the road.[14] Although it was now possible for the Mission to hire full-time workers, the scarcity of non-indigenous labour in the area made things even more difficult for the friars, who now had to send commissioners to recruit them from distant parts.[15] Montclar went even further and published the job opportunity in the press through encouraging letters describing the abundance of vacant lands and colonization opportunities across the region.[16]

In order to facilitate the labour recruiting and the organization of the works, the Capuchins devised a method that consisted of dividing the whole route into several sub-sections, and then assigning each of them to a 'caporal' (foreman). The caporal, meanwhile, was responsible for hiring a crew of 20 to 30 workers (Figure 2.3), arranging wages, and providing food – which was sold to him by a general supplier contracted by the Mission – and shelter if required (Gutiérrez 1921, p.344). The General Inspector or one of his friar assistants, on the other hand, supervised the works constantly, and sent periodical reports to the Minister of Publics Works in Bogotá and to the Governor of Nariño.

De Las Corts spent the first year working on the section from Sibundoy to Pasto, mostly reconstructing and finishing the trail the Mission had initiated years before. The most difficult part, as expected, would be the eight-kilometre stretch through the *páramo* of Bordoncillo some 3,200 metres above sea level. The cold temperatures, constant rains and remoteness forced the suspension of the works on various occasions. However, by the end of 1910, the friar submitted a very optimistic report. So far, summarized Fray de Las Corts, of the total 57 kilometres

Figure 2.3   Road crew, c.1910.
Source: Archivo Provincial de los Capuchinos de Cataluña.

composing this section, 33 had been completed, another 17 were
already contracted or under construction, and just about 6 (in the
páramo) were still to be contracted. The road dimensions, he added,
were the same at the top of the *cordillera* as in the valley: four metres
wide plus another 48 metres (24 metres on each side) of forest clearing.
Now that this part was nearly completed, he informed the Minister, his
plan was to delegate its supervision to another friar, while he would
devote his energies to the Sibundoy-Mocoa section so as to 'break as
soon as possible that wall that separates us from the capital of the
territory'.[17]

The 54 kilometres separating the town of San Francisco in the east-
ernmost part of the valley from the town of Mocoa proved to be a real
challenge for the stubborn friars. In January 1910, Fray de Las Corts
carried out a short expedition with the aim of studying the best route,
and reported to the Governor of Nariño that 'the ruggedness of the terrain, the innumerable creeks and inaccessible cliffs' had kept him and
ten workers wandering in the mountains for six and a half weeks.[18] Of
the two existent trails, the indigenous one described by Miguel Triana

Figure 2.4 Road crew demolishing a rock cliff, n.d.
Source: Archivo Provincial de los Capuchinos de Cataluña.

and the one opened decades before by the *caucheros* from Nariño, known as 'La Tortuga' (The Tortoise), the Capuchins chose the latter as the base route for the road. Although this was the shortest and less abrupt route, it had a considerable limitation: a significant portion of the road would have to be literally carved out of granite rocks (Figure 2.4).

The Capuchins needed three things to realize their dream of breaking the immemorial wall separating the 'civilized' and 'savage' worlds: additional funds, more workers and abundant dynamite. The first limitation would be unexpectedly solved thanks to a fateful event. By mid-1911, just when the budget initially requested by Montclar was exhausted and the missionaries were forced to fire 300 workers (of a total of 450) and take a loan to cover the debts,[19] the Peruvian army attacked a Colombian military garrison stationed in La Pedrera, a customs post on the Caquetá River. The incident, apart from fulfilling Montclar's prophesies, lent evidence to the obvious: that the country lacked accessible routes to its extensive and neglected Amazon frontier. A sad epitome of this drama, the defeated Colombian military expedition had spent four months reaching La Pedrera from Bogotá, having to descend the Magdalena River to the Caribbean, and from there navigate to the mouth of the Amazon and then upstream toward the Caquetá.[20] For the Capuchin monks, however, this event was like a blessing. On August 1911, barely a month after the battle of La Pedrera, the National Congress approved an additional budget of 36,000 pesos, and gave the explicit instruction of employing no less than 1,000 labourers on the works.[21]

With money flowing again and a paranoid government putting pressure from Bogotá by means of telegrams and letters to the Apostolic Prefect, the works soon acquired a frantic rhythm. Las Corts hurried to get tools, provisions and workers wherever he could find them. By the end of the year, an inspector sent by the Governor of Nariño reported that between Sibundoy and Mocoa there were currently 1,238 workers distributed in five sections, and also mentioned that in the previous months those came to exceed the 1,500 mark, producing an average of 25 workers per kilometre. The inspector, General Joaquín Escandón, who would later be involved in a fierce confrontation with Montclar, extolled the missionaries for the economy and organization of the works, stating that their involvement in the project was 'indisputably decisive in its success'.[22] Montclar, meanwhile, faced a hard time getting explosives. As soon as the funds had been approved, he had ordered 25 *quintales* (1,150 kilograms) of dynamite from Panama through the Ministry of Public Works. However, as the weeks passed and the precious material didn't arrive, the anxious friar went so far as to write directly to the President imploring him to send a telegram to Nariño telling the Governor to lend the Mission eight boxes of dynamite stored in Pasto.[23] The desperate measure might have been justified, for the explosives from Panama did not arrive until three months later, in February 1912.[24]

On 27th November 1911, Fray de Las Corts sent an 87-word telegram to Montclar announcing the near conclusion of the works and his triumphal entrance to Mocoa. 'Mocoanos unprecedented enthusiasm' – reads a fragment of the hasty note – 'authorities, missionaries, people received us with national flag, flower arcs, music, shotguns...willing spirit to plant flag and cross confines Colombian territory'.[25] Three days later, Gustavo Guerrero, Governor of Nariño, sent another telegram to the President confirming Fray de Las Corts' news and notifying the discovery of a 'rich marble mine' in the road area.[26] And on 1st December, Montclar, joyful, wrote a letter to the Minister of Public Works officially announcing the good news:

I offered the government to finish the road to Mocoa this year. I have the pleasure to inform your Excellency that the conclusion of the road is a fact as there are only two and a half leagues left and they will be most likely concluded before the year ends, because there are currently one thousand two hundred labourers working with extraordinary activity ... The missionaries, committed to the civilization of the Caquetá and Putumayo, to the prosperity and progress of that region, and seeking that Colombia, which gallantly protects the Missions, does not lose that vast territory, have worked tirelessly in the construction of a road that eases Colombia's maintenance of her rights over that region.[27]

According to an official report published by the end of 1912, it took the Capuchins six and a half months, from September 1911 to March 1912, to build the section from Minchoy (a few kilometres from San Francisco) to Mocoa (Figure 2.5).[28] Apart from the daily labour of about 1,500 indigenous and non-indigenous workers, the 50-kilometre-long, three-metre-wide road had consumed, according to Fray de Montclar, 3,220 kg of dynamite, not counting 'great quantities of ordinary explosives' (Figure 2.6).[29]

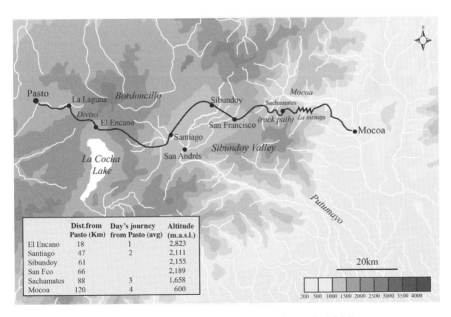

	Dist.from Pasto (Km)	Day's journey from Pasto (avg)	Altitude (m.a.s.l.)
El Encano	18	1	2,823
Santiago	47	2	2,111
Sibundoy	61		2,155
San Fco	66		2,189
Sachamates	88	3	1,658
Mocoa	120	4	600

Figure 2.5 The Capuchin road (section Pasto-Mocoa), 1912.
Source: Elaborated by author. Based on 'Croquis del camino de Pasto a Mocoa construido por los R.R.P.P. Capuchinos', 13th May 1912, AGN, MOP, Vol. 1408 (n.f.); and AGN, MOP, Vol.1415, fols.62–65.

Figure 2.6 The road, Sachamates-Cerreños section, n.d.
Source: Archivo Provincial de los Capuchinos de Cataluña.

Rituals of state-making

> It is in the realm of the symbolic production that the grip of the state is felt
> most powerfully (Bourdieu 1994, p.2).

The culmination of the 120-kilometre section from Pasto to Mocoa
filled the Capuchins with all sorts of eulogies and adulatory expressions on
the part of civil and military authorities who visited the new road. The
Treasury Prosecutor, for instance, submitted a detailed report to the
Minister of Public Works where he praised the economy and organization
of the works. He could not hide his amazement at the Sibundoy-Mocoa
section, and declared that 'never in Colombia had a road like that been
built, in less time, expenditure, and without fanfare' (Gutiérrez 1921,

p.340). Lucio Velasco, an Army Division Commander sent to Putumayo following the incident at La Pedrera, commented to the Minister of War that the road was of 'great importance to the army as well as for the country in general, in light of the current circumstances'.[30] The army officer was especially impressed by the road at the *páramo* of Bordoncillo, which he described as a 'titans' work', an epithet Montclar found particularly pleasing and would quote repeatedly.

Yet, no celebration of the Catalan missionaries' accomplishment can be compared with the inauguration event that took place in March 1912. We have a detailed account of this event thanks to the General Inspector of the road, General Joaquín Escandón, who wrote a journal describing the four-day journey of the official commission designated to receive the road.[31] Although the journal abounds in laudatory comments to the Capuchins, what captured Escandón's attention most, and to which he devoted more pages of his chronicle, were the exuberant receptions the commission was given in the towns along the road (Figure 2.7). The scene would be reproduced almost identically: first, the loud sound of rockets and the town's music band welcomed the delegation and announced its arrival, before they were directed to the *plaza* (central square); there, in the middle of palm arches and cheers to the government, the Mission, and the road, they were met by a parade composed of

Figure 2.7   Inauguration ceremony in Santiago, 10th March 1912.
Source: Archivo Provincial de los Capuchinos de Cataluña.

indigenous peoples (in La Laguna, Santiago and Sibundoy), *colonos* (San Francisco), or both (Mocoa). This enthusiastic greeting was followed by a scheduled programme, opened by the national anthem and flag-raising, and followed by solemn speeches on the part of the delegates, the missionaries and the local authorities.

The General Inspector described the speeches read by indigenous children from the Mission's schools as the most emotive acts of the ceremony. Probably written by the friars themselves but performed by a diligent pupil so as to offer the delegates an indisputable proof of the progress achieved by the Mission, their content was invariably the same: an exalted ode to the Mission in which the indigenous boy or girl praised the friars – predominantly the Apostolic Prefect – for having redeemed them from the state of 'savagery' and 'backwardness' that they inhabited until recently.

After four days of travel and many hours spent in the ceremonies organized by the Mission, the Commission finally arrived at Mocoa on the afternoon of 13th March. The inauguration event in Mocoa would be the most symbolic, not only because it was the conclusion of the journey but mainly because this place marked the point where the road culminated its strenuous passage through the Andes (Figure 2.8). There, as if the delegates were conquerors returning from a long heroic deed, apart from the music, rockets and cheers, they were greeted by a girl wearing a 'nymph' costume who offered them graceful crowns. The official ceremony took place on the 14th and was opened by a Mass officiated by Fray Fidel de Montclar, dubbed by Escandón as 'the alma mater of this redemptive enterprise [the road]'. Afterwards, with the crowd gathered in the plaza, the Apostolic Prefect gave a short speech in which he said he hoped the missionaries' accomplishment would be 'the precursor of an age of uninterrupted happiness and progress for the Nation', and with these words officially handed the road over to the Governor of Nariño and delegate of the President, General Gustavo Guerrero.

The Governor's speech, a short but dense talk overloaded with flattering superlatives for the Catalan missionaries – variously called 'titans', 'modern evangelizers' and 'sons of Colombia' – marked the culmination of the event. The entrenched *spatial* metaphor of the Andean 'undefeatable wall' was invoked again by the Governor so as to preface the friars' epopee, and its final defeat by 'pick, drill, and dynamite' to signal the *temporal* passage from savagery to civilization and progress; and so the road, though an inert stretch of dirt, came to life in the speech as a living infrastructure driving the pace of change, for, in his own words, 'it makes a steady way to trade, agriculture, industry and, in short, to the movement of an enormous wealth, stagnant for centuries'. When the Governor had concluded and stepped down from the platform improvised for the occasion, recounts Escandón, he was greeted with the warm ovation of

Figure 2.8 Inauguration ceremony in Mocoa, 14th March 1912 (Delegates shown at centre).
Source: Archivo de la Diócesis de Sibundoy.

the attendees, the national anthem and a rifle salute by the guard of honour. Finally, after two more brief interventions by Fray de Las Corts and Rogerio Becerra, the civil Prefect, the inauguration was officially closed at noon, though, notes the Inspector, the delegates continued receiving manifestations of gratitude during the rest of the day while the 'populace' enjoyed a beef meal offered by the Mission and the government.

Figure 2.9 shows the Commission entering the town of San Francisco, the town funded by the Capuchins in 1902. The delegates (on horseback, from left to right: General Gustavo Guerrero, Fray Fidel de Montclar, Fray Estanislao de Las Corts, General Joaquín Escandón) listen attentively to a speech read by a peasant boy (barefoot, centre), while the town's musical band, other children holding palm branches and the national flag (right), and some spectators (background) witness the solemn event. This particular moment is not recorded in Escandón's journal, and the picture comes from a different source. Yet, if we consider that the different reception ceremonies had been carefully planned by the missionaries and were replicated in different places, one can imagine this same picture and the others displayed at various places during the Commission's journey.

I would like to stress both the solemnity conveyed in the picture and the replication of the moment it captures, for it is here that one of the

Figure 2.9 'Triumphal entrance of the Governor and the Apostolic Prefect to San Francisco', 12th March 1912.
Source: Archivo Provincial de los Capuchinos de Cataluña.

central features of the inauguration, namely its *ritualistic character*, is plainly revealed. The ritual, moreover, although revolving around a single episode – the inauguration of the road – was by no means restricted to it. For, as noted by Montclar, the road was largely conceived as the 'precursor' to a 'radical transformation' or, as he depicted it in his lecture at the Faenza Theatre, as a landmark signalling a transcendental break in the history of the Colombian Amazon. This was precisely what the whole event was about: not so much to celebrate the road itself, but to expose and perform, through all the solemnity in which the ritual was embedded, its manifold *effects* on this frontier territory.

We might deconstruct the images and the Inspector's chronicle in order to dissect the whole ritual into discernible parts, thus identifying the meanings embodied in the different symbols and elements – the cross, the flag, the anthems, the parades, the protocols followed – that together mirror an important dimension of state-building. Still, rather than emphasizing how the state is produced and enacted through the performative nature of the ritual, I would like to examine the ritual in its capacity to render visible the myth in which the state is rooted. Therefore, at this point we must ask why, *through* ritual, does a lifeless infrastructure metamorphose into a living actor over-saturated with nationalistic, religious and economic meaning? The answer, I suggest, lies not so much in the omnipresent image of the road as marking the 'defeat' of the granite 'wall' that kept the Putumayo – with all its buried treasures and souls to conquer – isolated for centuries, as in what this image conjures up, and what is ultimately ritualized in the inauguration event: the emancipation of *man* from *nature*.

Let me quote two passages related to the road that powerfully invoke this act of emancipation. The first is an excerpt from a chronicle published in 'La Sociedad', a newspaper from Bogotá, which describes the magnitude of the Capuchin's enterprise as seen through the eyes of a 'civilized' inhabitant from the country's highlands. The second is a fragment from Escandón's journal where he narrates the 'colossal duel with nature' unleashed by the Apostolic Prefect on the San Francisco-Mocoa section.

> Imagine a spot in the most precipitous and abrupt part of our Andes, in a long canyon formed by two steep cordilleras, and in the background, meandering, the Mocoa River. Detonations echoing from cliff to cliff reach our ears, making the effect of powerful artillery batteries in the midst of a tough battle. One thousand and six hundred men, deployed in numerous guerrillas and filling an extension of ten kilometres, give the impression of a hardened army attempting to defeat the enemy on its way. *The illusion is complete.*

> … In reality, one thousand six hundred workers, divided in a multitude of crews, provided of enough tools, commanded by the reverend Capuchin fathers, attempt to break the wall, apparently insuperable, that hindered the access to the Amazon world. Massive and imposing rocks blow into pieces by the power of dynamite, which handled by skilled officers, ravages those crags that had defied the centuries (Las Misiones en Colombia, 1912, pp.118–119, emphasis added).

> [Montclar] undertook [the duel] with vigour, he fought, and he defeated: the upstanding jungle bowed to him, while the crags and mounts stroked his feet, and the waterfalls, rivers and abysses made way for him; nature greeted him reverently, acknowledging him as his legitimate conqueror and victor in fair combat; and the undaunted son of Saint Francis of Assisi entered Mocoa, holding in his right hand the crucifix and on his left the level and the compass.[32]

What we witness in this celebration of the violent conquest of nature by man, so overwhelmingly present in the visual and literary rhetoric of the inauguration rituals, is the reification of the classical myth through which an untamed nature is subdued so it can give way, to quote Hegel, to a 'second nature' (the state) embodying the realm of reason (Hegel 1976, p.285).[33] This same myth, moreover, is continuously evoked in the

Figure 2.10   Dynamiting the mountain, Vijagual creek, c.1911.
Source: Archivo Provincial de los Capuchinos de Cataluña.

various pictures of workers dynamiting and hammering granite rocks
under the command of a few feverish monks determined to defeat the
inhospitable nature of the *cordillera* (Figure 2.10). This is another kind
of ritual, one that exalts the creative destructive power of the road (the
'illusion' it creates, as the chronicler tells us) through the very act of
breaking the immemorial 'Andean barrier' separating the 'savage' and
'civilized' worlds.

    The main effect of this creative destructive process is that, *through
ritual*, all the violence it embodies, like the granite stones of the *cordillera*, is normalized or simply erased from the landscape. So powerful is
the effect of ritual that, to quote Taussig's remark on the essence of state
fetishism, 'the signifier is itself prized apart from its signification so as to
create a quite different architecture of the sign – an architecture in which
the signified is erased' (Taussig 1992, p.128). And what do the signifier
and signified stand for in this ritual? The former naturally alludes to
the state, the 'second nature' incarnated in the Apostolic Prefect and his
'artillery battery' of road workers. It is the *creative* part of the process, the
artifice whereby the state comes to appear as an abstract force detached
from and standing *above* society and nature. As for the latter, it denotes
the *destructive* part of the process or what needs to be annihilated

(the 'first nature' embodied in the savagery of the Indians and the Dantesque topography of the *cordillera*), and the *means* this act of annihilation requires: labour, dynamite, the crucifix.

* * *

In the following chapters, I shall show how, at the level of *practice*, neither the state retains the immanent power granted at the level of *myth* nor are society and nature just passive agents shaped by domination or governmental practices. Certainly, the history of the road is not exactly a story of man's unassailable victory over the untamed wilderness, and neither is it a story of the state's relentless expansion over the 'savage' frontier. However, and by way of conclusion, I want to lay emphasis here on the significance of ritual within the context of the road, as it represents an essential element in the very process of state-building.

As noted, it is through ritual that the violence implied in the creative destructive process of the road is normalized or rendered legitimate. The most manifest form of violence in the early history of the road relates to the hundreds of indigenous peoples forced to work on it. There exists an old legend among the Inga and Kamëntza indigenous peoples from the Sibundoy Valley associated with this episode. First collected by Fray Jacinto de Quito (1938), the legend, known as 'La Leyenda de Nuestro Señor de Sibundoy' [The Legend of Our Lord of Sibundoy], goes as follows:

A hunter had been living for a long time at a place known as 'The Cedar' near the San Pedro River. One morning, he put poison on his arrows, and having put them in his quiver, took his bow and set off hunting in the direction of Sibundoy. He had been walking for some minutes, when he saw a huge bird, rather like a condor, flying very fast. He watched to see where it would land, and it alighted on a myrtle. Since the undergrowth around the tree was very thick, the hunter had to bend low, and force a way through with his hands, thus approaching his prey rather as a wolf would. But then he stood up in front of the myrtle … he remained motionless, paralysed by what he saw: at the foot of the tree, in a hole shaped like a niche, Our Lord of Sibundoy was sitting barefoot, his hair long, clothed in 'cusma' and 'capisayo'.[34]

'I want you', said the Lord to the hunter, 'to call the Indian Chapter so I can tell them what they have to do with me'.

The hunter, without uttering a word, ran to follow the order. Soon, not only the Indian Chapter but the whole town gathered at the place of the apparition.

'I come to live among you, he said to them, but on condition that you obey me and abandon your bad customs. And I want you to build a church in this place'.

The Governor, without answering him, addressed his people and commanded them to start working on the Church. Some went to gather wood, others vines and palm leaf, etc. And soon the Church was finished. Then, the same Governor appointed sacristans to take care of the Lord.[35]

This part of the story – which according to Jacinto de Quito constitutes the basis of the old legend – was expanded as follows as a result of the building of the road:

The sacristan, noticing that the Lord's clothes were wet through every morning, suspected him of going out at night, and decided to find out. Under the pretext of renewing the candles, he went to the church at late night...

The Lord had vanished. He told the Governor, who, in agreement with the other leading men, ordered that the culprit be beaten with twelve strokes of the whip. After this punishment had been carried out, the Indians expected that the Lord would ask for forgiveness and promise not to do it again; but instead he stood up, turned his back upon them and went off on the road to Pasto ... They tried to catch him but he became invisible, and all they could do was go back to the village sadly, and repent of having beaten him. Their remorse became even more acute when they discovered that the reason why the Lord had gone out each night was to take their place working on the road from Pasto to Mocoa, which he did in the following way:

When night approached, he grabbed a machete and a candle and went to work on the road and to build bridges over the rivers. He stopped only at daybreak, when he came back to the church and rested in his niche with his clothes wet.

The Indians, when travelling through the new road, upon finding it blocked by palisades or slides, followed the wax tears left by the candles to avoid getting lost, and this way they walked safely.

In Fray de Quito's narrative, this legend has the purpose of illustrating the 'superstitious' character of the natives, and hence the 'terrible struggle against the law of custom' undertaken by the missionaries (Quito 1938, p.32). The ethnologist Juan Friede has another hypothesis. According to him, this legend is about how the indigenous peoples from Sibundoy 'transform in their minds those historic events and those social problems which they are not able to face in practice' (Friede 1945, p.318).

It is not the purpose here to analyse this legend in detail or its significance in the social and cultural context of the indigenous people from the Sibundoy Valley. Yet, I would like to suggest that it is revealing of how the Inga and Kamëntza struggle to make sense of the violence they were subjected to during the construction of the road. This violence, we know, was inflicted on the Indians themselves but appears here projected onto the image of the Lord of Sibundoy; in other words, as noted by Friede, this constitutes a violence that cannot be faced directly. Furthermore, once projected, this violence seems to reflect back upon the Indians in the form of guilt (for having unjustly punished the Lord of Sibundoy, who was working in the road for their sake), with the effect that at the end this same violence (the one imposed on the Indians) is normalized or concealed.

This is precisely the same effect that we encounter in the road's rituals through their re-enactment of the state myth. To borrow Abrams' expression on state fetishism, through these rituals the state is revealed as 'the mask which prevents our seeing political practice as it is' (Abrams 2006, p.125). The crucial aspect of such rituals, however, is not the way in which they conceal the state practices behind a mask (e.g. the creative destructive process through which the Indians were dispossessed of their labour and lands) but how they reveal this same mask as constitutive of the state. Put another way, the significance of ritual lies here not in exposing the fiction of the state (its 'civilizing' mask) but in how it endows this fiction with an appearance of reality. This appearance or illusion of reality, seen as a practice rather than a mask of the state, is essential to grasp the violence involved in the history of the road and how it was rendered invisible or legitimized. And it is also essential to comprehend the different dynamics that the road assisted and how through them, or even in spite of them, the state revealed its hegemonic character, the subject of the next chapter.

Notes

1   The Apostolic Prefecture of the Putumayo and Caquetá, which at the time of its creation (1904) encompassed an extensive area (around 131,940 square kilometres), comprised a part of the old *Territorio del Caquetá*, which gradually split into administrative units such as *Comisarías especiales* (Special commissionerships), *Intendencias* (Intendancies) and finally *Departamentos* (Departments). For a detailed historical account of the administrative and territorial changes of the Putumayo and Caquetá territories in the nineteenth and twentieth centuries see Codazzi (1996, pp.54–57), Ramírez (2003, pp.203–239) and Vilanova (1947, vol.2, pp.131–139).

2   For a description of the government regulations and agreements pertaining to the Catholic Missions in the context of the conservative hegemony see Vilanova (1947, pp.103–115) and Bonilla (1972, pp.52–60).

3   'Informe sobre camino de Nariño al Puerto de La Sofía', 21st August 1906, AGN, MOP, Vol.1407, fols.30–37v.

4   The Putumayo remained under the jurisdiction of the department of Nariño until 1912, the year that it was established as a *Comisaría especial*, dependent directly on the central government.

5   Bucheli to MOP, 7th September 1906, AGN, MOP, Vol.1407, fols.28–29v.
6   'Decreto número 302 de 1905', 27th December 1905, AGN, MOP, Vol.1407, fols.8–10v.
7   Reyes to Bucheli, November 1906, AGN, MOP, Vol.1407, fols.39–40, 70–72, 63–66.
8   'Contrato', 25th February 1907, AGN, MOP, Vol.1407, fol.100v.
9   'Informe de Víctor Triana sobre el trazado de la vía entre Pasto y el Putumayo', May 1908, AGN, MOP, Vol.1407, fols.173–177v.
10 Reyes to Bucheli, 9th January 1908, AGN, MOP, Vol.1407, fol.133.
11 'Acuerdo Presidencial XVIII', 26th May 1908, AGN, MOP, Vol.1407, fol.154.
12 Fray de Montclar to Minister of Government, 2nd September 1909, AGN, MOP, Vol.1407, fols.194–198v.
13 The complete pamphlet can be found in the Mission's annual report of 1912 (*Las Misiones en Colombia* 1912, pp.115–117).
14 'Resolución No.264. Gobernación Departamento. Pasto, 17 Nov. 1909', APCC, Missions Caquetá-Putumayo, Box 4, n.f.
15 De Las Corts to Minister of Public Works, 22nd January 1910, AGN, MOP, Vol.1407, fols.262–264.
16 'Camino de Oriente', *La Renovación* (No. 309, Bogotá), 9th January 1911.
17 'Informe de los trabajos del camino del Oriente', 30th October 1910, AGN, MOP, Vol.1407, fols.323–328v.
18 'Informe de los trabajos del camino de Oriente al gobernador del Departamento', March 1910, AGN, MOP, Vol.1407, fols.290–292v.
19 AGN, MOP, Vol.1407, fol.410.
20 See Forero (1928) for or a detailed account of the La Pedrera military expedition.
21 AGN, MOP, Vol.1407, fol.414; 'Camino del Putumayo. Su pronta terminación', 18th October 1911, *Gaceta Republicana* No.676, Bogotá, in APCC, Missions Caquetá-Putumayo, Press, fol.87.
22 Escandón to Secretary of Treasure (Nariño), 31st December 1911, ADS, folder 11-10-01, Camino Pasto a Puerto Asís, n.f.
23 Fray de Montclar to President, 26th November 1911, AGN, MOP, Vol.1407, fols.461–462.
24 Fray de Montclar to Minister of Public Works, 23rd February 1912, AGN, MOP, Vol.1407, fol.507.
25 Fray de Las Corts to Fray de Montclar, 27th November 1911, AGN, MOP, Vol.1407, fol.458–459.
26 'Camino a Mocoa', *El Nuevo Tiempo*, 30th November 1911, No.3221, in APCC, Missions Caquetá-Putumayo, Press, n.f.
27 Fray de Montclar to Minister of Public Works, 1st December 1911, AGN, MOP, Vol.1407, fol.467.
28 *Camino de Pasto a Puerto Asís. Relación de viaje. Informe de la Comisión nombrada por el Gobierno Departamental de Nariño para inspeccionar la vía*, Imprenta del Departamento, Pasto, 1912, in APCC, Missions Caquetá-Putumayo, Box 3, n.f.

29 Fray de Montclar to Minister of Public Works, 20th February 1912, AGN, MOP, Vol.1407, fol.526v.

30 Velasco to Minister of War, 13th September 1912, AGN, MRE, Diplomática y Consular, Box 742, folder 323, fols.27–29v.

31 *Informe sobre la terminación del Camino de Mocoa*, Imprenta del Departamento, Pasto, 1912, in APCC, Missions Caquetá-Putumayo, Miscellany, n.f.

32 *Informe sobre la terminación del Camino*, 16–17.

33 This myth is a prominent theme among social contract theoreticians such as Hobbes, Locke, Montesquieu and Rousseau, to whom notions of peace, security or freedom – cornerstone elements of the modern state – inexorably stem from a distinction between the state of nature – ruled by instincts and passions – and the civil state – governed by reason (see Hobbes 1937; Locke 1980; Montesquieu 1990; Rousseau 2004).

34 The *cusma* and *capisayo* form part of the Sibundoy indigenous peoples' traditional costume. The former is a kind of sleeveless narrow robe made of cotton, while the latter is a long and wide *ruana* or poncho.

35 The version reproduced here corresponds to the abridged English translation of the original (included in Quito's Memoirs; see Quito 1938, pp.89–92) published in Bonilla (1972, pp.91–92). I have translated and added some fragments to the original that were omitted in Bonilla's book.

3

Fray Fidel de Montclar's deed

The completion of the road from Pasto to Mocoa holds a special signifi-cance in the history of the Putumayo. Historically, this event signalled the culmination of long-standing dreams, plans and efforts to overcome the geographical barrier separating this territory from the rest of the country. Symbolically, it powerfully evoked and reified the image of the state as an all-embracing 'civilizing' power inexorably expanding over the 'savage' frontier. The significance of this image, as has been argued, lies not in how real or fictitious it was but in how through the illusion of reality it projected, the rhetorical and physical violence it embodied remained con-cealed or was rendered legitimate. Some of the dynamics involved in the construction of the road, from the dispossession of indigenous labour to the dynamiting of the Andean cordillera, represent visible examples of this violence. But the history of the road went far beyond these episodes and did not end with its imposing inauguration event of March 1912. This is so not only because the road originally aimed to reach a navigable point in the Putumayo River (Mocoa was just halfway to that point), but because it was envisaged as an essential infrastructure to assist other state policies and plans. These included fostering the colonization of the Putumayo's 'vacant lands', stimulating commercial activities across the Amazon region, providing access to the Atlantic Ocean via the Putumayo and Amazon rivers, reinforcing the military presence in the national bor-ders, and aiding the expansion of the Capuchin Mission's 'civilizing work'. Together, those plans and the ways in which they were conceived and translated into practice reflect the policies and practices of frontier-making during the first decades of the twentieth century.

Frontier Road: Power, History, and the Everyday State in the Colombian Amazon,
First Edition. Simón Uribe.

This chapter revolves around these policies and practices, which I seek to explore and interrogate through the different characters, conflicts and events involved in the building of the road, as well as some of the broader social and spatial dynamics this infrastructure assisted. In doing so, I will further delve into the question of hegemony, and specifically with the ways in which this concept allows us to have a better comprehension of the relationship of inclusive exclusion between frontier and state.

An important dimension of hegemony has to do with, or at least rests – as in Weber's notion of the state – upon the monopoly of coercion (Weber 1998, p.78), a situation that in practice, more seldom than not, stems from consent. The history of the road can, at least in part, be read from this purely coercive dimension, and more explicitly as a creative destructive process through which the state violently attempted to conquer and expand into 'savage' territories and populations. As previously emphasized, this process cannot be detached from the nineteenth-century *criollo* imaginary order of the nation, an order whose hegemonic condition was largely edified through and depended on the production and perpetuation of racial and spatial dichotomies. Those dichotomies, as well as the characters, representational forms and rituals through which they were constantly conjured and reproduced, are not only over-whelmingly present in the history of the road but are essential to grasp the relationship between hegemony and violence.

But the essential aspect of hegemony, in its Gramscian sense, is that it helps understand power in relational rather than linear or absolute terms.[1] This means that power is not a zero-sum game but an unstable and con-tested field where multiple forces and interests converge. This is to a large extent the case of the road, if we consider the indefinite everyday dramas and conflicts involved in its conception and construction. This does not mean that the road and the different dynamics linked to it did not entail significant historical and spatial transformations. Nor does it imply that the vision inspiring this infrastructure simply faded away when confronted with the harsh obstacles it faced on the ground. Rather, what this story unveils is precisely the recurrent gap between the power of the state as *projected* at the level of myth and *realized* at the level of practice.

The abundant archive records documenting the countless difficulties and conflicts present in the history of the road clearly evidence this gap. Some of these difficulties and conflicts include endless bureaucratic obsta-cles frequently bringing the road works to a halt, everyday disputes with road workers and users, and harsh confrontations between the Capuchins and their political and personal adversaries. Another element that occupies a central place within this history is the role played by nature, an actor whose agency is often ignored or under-emphasized and yet sheds impor-tant light on the unstable character of state policies and practices.[2]

How, despite instability and conflict, can we still think of hegemony as a useful concept to approach state-building processes (or, conversely, to talk of the state in terms of a hegemonic order)? The answer to this question, as implied in Gramsci's analysis of power, is precisely that hegemony enables us to conceive domination as a process that is not external to but coexists with struggle and conflict. This point brings us back to the argument introduced in the previous chapters about the importance of conceiving the state as a force that is simultaneously ideological and material or, to quote Fernando Coronil, as a *form* that 'works by establishing a relationship of equivalence between the general and the particular, the abstract and the concrete' (Coronil 1997, p.116).

The road's documents examined in this chapter (official reports, letters, telegrams, pictures, 'cadastral' plans and maps) shed light on both the abstract and concrete dimensions of the state and state violence: they speak of the 'civilizing' ideology and logic driving the project of the road; of the many everyday dramas, power disputes and ceaseless problems and obstacles 'on the ground'; of the different and often contested ways through which the road was conceived and appropriated; of the language defining the terms and boundaries within which struggles and conflicts were expressed; and of the implicit and explicit silences and erasures present in this language. Again, the fundamental aspect of these documents and the different situations they describe is not what they tell us about one or another dimension of the state, but how they mirror the different ways in which such dimensions are mutually interdependent. And it is precisely this interrelation between the abstract and concrete or discursive and material, I argue, that allows us to understand how hegemony works.

Burdens and loads

For the native peoples of the Sibundoy Valley and the Putumayo foothills, who for centuries had remained relatively isolated from the 'civilized' side of the country, the defeat of the 'Andes barrier' by the hands of the Catalan missionaries constituted a dramatic event. The burdens, humiliations and violent aggressions to which they were continuously subjected from the early days of the road, would soon make clear who would bear the brunt of this infrastructure. One of the main paradoxes of this episode, as the legend of Our Lord of Sibundoy referred in the previous chapter suggests, is how the Indians, aided by the Lord, were to work on the missionaries' project as if it was meant to be their way to redemption.

It is difficult to establish the number of indigenous labourers involved in the road's construction or how much of this work was compulsory or unpaid. The Mission's reports regularly indicated the number of workers and wages paid, but they did not discriminate between indigenous and non-indigenous labour. The fact that as early as 1909 the missionaries were forbidden to impose labour obligations on the native population might help explain why, if they intended to continue forcing the Indians to work on the road, they had no intention to report it. On the other hand, this prohibition was only partial. The resolution issued on this subject excepted work classified as 'punishment' under the 'Reglamento para el Gobierno de indígenas' (Indigenous' Government Code).[3] The Reglamento, a sort of constitution drafted by Fidel de Montclar for the indigenous communities under the Mission's jurisdiction, contemplated various crimes such as thievery, drunkenness, public meetings and parties, and the penalties established included compulsory work from 1 to 20 days. In any case, and regardless of the extent to which this code constituted a regular source of indigenous labour, there is plain evidence that at least during its early years the Indians were frequently compelled to work on the road for no pay.

As formerly noted, the Capuchins not only began to build the road with only the support of the Indians' tributary work, but had actually lobbied for the Sibundoy route, arguing the abundance of free labour in the area. In September 1909, the central government approved the 40,000 pesos budget requested by Montclar, which allowed the Mission to hire paid labour, but the pressure on the Indians did not decrease. Two months later, the indigenous authorities from the Sibundoy Valley presented a memorandum to the Governor of Pasto complaining about this situation. They claimed that each of them was compelled to supply the Mission with 100 or more workers per week, and protested about the harsh treatment they were given.[4] Although these claims were the basis of the aforementioned resolution, this legal measure would have limited application. This is illustrated not only by the fact that the missionaries could still compel Indians to work on the road as a form of punishment, but that indigenous claims on this regard did not cease.[5]

The scarcity of non-indigenous labour in the Putumayo by the time the Mission was commissioned to open the road also meant that, for various years, the Indians did the bulk of the waged work. In October 1910, for example, when the works were in full swing and the number of workers approached 1,500, Las Corts reported to the Minister of Public Works that 90% of them came from the indigenous town of Santiago.[6] Even though this proportion would gradually change as a wave of *colonos* began to arrive and settle in the region, the Indians continued working on the road and often performed the most burdensome tasks. This was the case of the *cargueros* (porters), a strenuous chore carried out equally by men and women, and often by children. In 1913, for instance, an inspector sent by the central government to oversee the works reported that 7 out of the 34 workers currently employed in the Sibundoy Valley area were children aged between 10 and 13. Upon questioning a foreman about this subject, the inspector pointed out that his answer was simply that children 'worked better' than adults in the loading of construction material.[7]

The *cargueros* were more commonly employed in the transport of supplies demanded by the missionaries and the road crews. A detailed invoice sent to Las Corts in September 1912 by the road general supplier in Pasto provides us with some details about this trade.[8] The bill listed the names of porters dispatched from Pasto between 7th August and 18th September and included the load carried by, and wages paid, to each one. With the exception of two males – whose high loads assigned suggest that they had pack animals – the 71 porters listed bear indigenous surnames, mostly Inga and Kamëntza, and the proportion of men and women is almost equal. As the list does not provide ages it is not possible to establish the number of children-porters, although their presence is indicated in various cases where the person is registered as '[name] son of [name]'. The loads oscillate between one and two-and-a-half 'arrobas' per person (25 and 62 pounds approx.), and the freight's prices between one and two silver pesos, a variation probably indicating the load destination, which could usually vary in time from 2–7 days. The poor wages paid for this onerous job are hinted if we contrast them against the prices (at the Pasto marketplace) of some of the basic goods listed in the same bill: 30 cents for a knife; 80 cents for a pound of bread and 60 cents for butter; 3 pesos for a pair of trousers and 35 pesos for a barrel of wine.

The burdens borne by the indigenous carriers were not confined to the long journeys, the low wages, and the excessive loads which, as referred to previously, also included carrying humans. Abuses and mistreatments on the part of the road crews stationed along the way were not uncommon. In November 1912, for instance, Rogerio Becerra, Mayor of Mocoa, wrote the following missive to all the foremen drawing their attention to the recurrent cases of thievery and aggression against the carriers:

> The workers labouring on the road and the telegraph cause great harm to the Indians, sometimes spoiling them, others treating them like beasts and stealing the rubber they carry to Pasto, thus making serious damage since the rubber owners have them [the Indians] arrested and charge them with the quantity stolen.[9]

These aggressions continued to persist and the Indians sporadically responded to them, as suggested by events such as the shooting of a horse belonging to the road by the Sibundoy Indians reported in May 1914.[10]

Yet, although highly symbolic of the daily burdens and violence born by the natives on the road's construction, these events constituted the prologue to more complex and lasting dynamics that this infrastructure promoted and assisted. Those dynamics, some of which are described below through the history of two different towns founded by the Capuchins, expose the violent politics and practices of state-making in the Putumayo.

A tale of two towns

As noted, Fray Fidel de Montclar, head of the Mission and the main supporter of the road, always conceived this infrastructure project not as an end in itself but as the precursor to a 'radical transformation' of the approximately 130,000 square kilometres he was entrusted with as Mission territory. This transformation stemmed from and was grounded on three central pillars, which largely summarized the practices of statecraft in the Amazon during the first decades of the twentieth century: civilization, material wealth and sovereignty. The Mission's reports submitted annually to the central government provide abundant evidence of the progress achieved by the Catalan friars on those fronts. As these reports usually include long statistical annexes containing a wide range of information, from towns and schools founded to baptisms and weddings celebrated, it is possible to follow those variables annually. This task, however, is facilitated by the friars themselves. Eager to show the progress of the territory entrusted to the Mission and defend themselves from the frequent attacks by their political adversaries, the missionaries sporadically published leaflets and brochures containing 'past vs. present' accounts of this territory.

A good example of this missionary propaganda material is found in Fray Pacífico de Vilanova's extensive account of the first 25 years of the Mission in the Putumayo and Caquetá. According to the Capuchin friar, in 1906, a year after Fray de Montclar had been appointed Apostolic Prefect, the Indians were in a 'pitiful state, lacking connection with civilized elements, except for the places dominated by *caucheros*'; the mission territory only had the town of Mocoa and a 'few hamlets,' 'two or three thatched chapels' and five schools. For the year 1930, he referred to the following numbers so as to allow readers to 'judge for themselves' the transformation of the large territory under the tutelage of the Capuchins: about 19,700 'civilized' *colono* settlers established in the Mission territory, 6,600 Indians in the process of being 'reduced to civilized life', and around 6,300 'savages' still to be converted to the Catholic faith; 39 towns established by the Mission and another four under construction; 32 churches, 62 schools (half of them built by the Mission), two orphanages, a hospital and five dispensaries; and several trails, bridges and navigation routes connecting different points of the Apostolic Prefecture, all built by the Capuchins. As for the road, its prominence within the Capuchin's transformative quests is stressed in the friar's claim that 'it has changed the region's face entirely' (Vilanova 1947, vol.2, pp.273–277).

It is not my aim here to discuss at length the historical legacy of the Capuchin Mission, and this constitutes a subject that has been widely addressed elsewhere.[11] However, and for the sake of the argument of this chapter, I will consider the history of two towns, Puerto Asís and Sucre, which are illustrative of the process of state-building in the Putumayo. Both histories, moreover, shed important light on the particular dynamics that the road enabled or assisted and, at a broader level, on the central role that this infrastructure played within the spatial and historical configuration of the frontier.

Puerto Asís

The idea of Puerto Asís was born with the road. As early as 1906, Miguel Triana, the engineer sent to the Putumayo by Reyes to study a route from Pasto to La Sofía, mentions that Estanislao de Las Corts already had the project in mind. The Capuchin friar, according to Triana, had strongly argued in favour of building the road through the Sibundoy Valley and Mocoa, citing the abundance of free labour, and from there to a point on the Putumayo River; there, added Las Corts, they planned to found the 'City of Asís', a town conceived as the heart of a 'great vacant region' to be opened up by the road (Triana 1950, p.352).

Five years after Triana met the enthusiastic friar, the latter was directing the road construction works on the Pasto-Mocoa section and Fray de Montclar was lobbying in Bogotá in order to make the plan of Puerto Asís a reality. In July 1911, he presented a colonization project to the National Congress. The project, a three-page typescript text, basically outlined the basis of what the Apostolic Prefect deemed to be the imperative steps the government should follow to avoid losing the 'immense and lush jungles of the Caquetá and Putumayo' into the hands of Peruvians.[12] First of all, the roads from Guadalupe to Florencia (province of Caquetá) and Pasto to Mocoa (Putumayo) should be completed without delay, and then extended to a navigable point on the Orteguaza River (tributary of the Caquetá River) and Putumayo rivers, respectively. Subsequently, five or ten square kilometres of land around each of those points were to be cleared, the town plan drawn, and the first *ranchos* (huts) built. Upon conclusion of this task, each settler family would be allotted a *rancho* and a 20-hectare land plot, in addition to tools and seeds, and provided with food for the first six months. The settler family should remain in the colony for at least two years in order to be granted property, or otherwise the house and land plot would be re-allocated to another *colono*. The project was designed for poor and landless peasants from other parts of the country and did not grant money to the settlers, with the excuse that they would spend it and then abandon the place. The total budget required for the project, initially conceived for 100 families, was estimated by Fray de Montclar as being 52,000 pesos. Finally, the friar contemplated a second stage with foreign migrants, and recommended a pilot project with 25 families, all of whom should be 'moral and catholic individuals' so as to avoid 'dangerous persons of depraved habits and anarchic ideas'.

The congressmen, most of whom only knew the Colombian Amazon through maps and feared the loss of this vast territory (the siege of La Pedrera had recently taken place), eagerly embraced the Apostolic Prefect's initiative. Las Corts assumed the task of locating a suitable place for the colony, and eventually chose a spot on the left margin of the Putumayo River, not far from Reyes' port La Sofía and about 90 kilometres south of Mocoa. There, through a Mass officiated by the same friar on 3rd May 1912, Puerto Asís officially came to life.[13]

In the Mission's annual report of that year, Montclar devoted several pages to praise the privileged location of the colony. In the first place, he stressed its strategic military position by providing a chart of time distances to different points across the region in dispute with Peru, and gave an assurance that from there Colombia could secure its rights against both Peru and Ecuador. Secondly, and resuscitating Reyes' inter-oceanic project, he emphasized its commercial advantages thanks to its fluvial access to Europe and the United States via the Putumayo and Amazon rivers. Last but not least, the friar considered the significant benefit the new foundation represented for them, by noting how the town would be 'the point of departure from where they will leave to announce the faith to the wretched sons of the jungle' (*Las Misiones en Colombia* 1912, p.134). Montclar's arguments, all imbued with a spatial content, are revealing of the friar's geopolitical vision of the state.

Las Corts, now head of the Colony, had begun the first works by mid-1912 with the help of the first settlers and some indigenous families from the surrounding area. A few hectares of forest were cleared, and a provisional church as well as a big two-story house for 100 *colonos* were built. Soon, however, they were faced with the problem of provision shortages. The most basic supplies such as food, tools and clothing had to be brought on the backs of Indian carriers from the city of Pasto more than 200 kilometres away, involving several days of travel. Even considering the meagre wages paid to the Indians, Fray de Las Corts complained that the transport costs were 'much higher' than the value of the goods themselves (*Misiones Católicas del Putumayo* 1914, p.45). In order to ease the problem of food supply, the Colony's most pressing issue, the energetic friar rushed to start the plantation work. However, facing a scarcity of settlers, he wrote on 7th May to Montclar that he had 'forced' the [indigenous] villages of Guamués, San Diego and Ocano to plant five *cuadras*[14] each'.[15] In that year's official report, Montclar would subtly change this version declaring instead that Indians had been 'invited' by the missionaries, and that even those living three days from Puerto Asís had come to help with their families for several days (*Las Misiones en Colombia* 1912, pp.135–136).

The early lack of settlers was largely related to the fears among *colonos* about the Puerto Asís climate, a situation the Apostolic Prefect attributed to a campaign on the part the 'Mission's enemies' (*Las Misiones en Colombia* 1912, p.134). To overcome this impasse, and anxious to show results to the government, as early as April 1912 the Apostolic Prefect had commanded Fray Andrés de Cardona, in charge at that time of the road works from Mocoa to Puerto Asís, to recruit 200 labourers and send them to the colony at once. In order to persuade them, Montclar asked him to make them all sorts of promises including higher wages than those paid on the road; those reticent to go, conversely, were to be threatened by assuring them they would not be hired by the Mission again. 'At any rate', he urged Cardona:

> It is good to have quite a few people going to Puerto Asís because in Bogotá they are persuaded that the Colony is well advanced, and it would be a tremendous discredit if after so much talk and fanfare we end up having nothing.[16]

The Apostolic Prefect visited Puerto Asís in July and, perhaps trying to downplay the various problems currently afflicting the colony, showed himself to be highly optimistic about the progress achieved so far. He reported that there were 20,000 plantain shrubs, 30 hectares of sugar cane and great quantities of corn, manioc and rice planted. He found three new houses and, excited by the arrival of new settlers, ordered the construction of another 30. As for the authority of the missionaries among the villagers, he observed that 'the missionary is for them all what in a well-organised society the civil, ecclesiastic, military, and political authority can be; in other words, there was no other authority than the Priest'.[17]

In practice and 'unofficially', however, things didn't look that bright. The arrival in early September of 1912 of a military contingent of 500 men, together with the harsh winter that hit the Putumayo that year, left the provision of supplies in a critical state. Desperate, Las Corts sent a letter to Montclar grumbling about the situation:

> What can I do in the Colony with so many people, without dishes, cups, spoons, etc. etc.?
>
> I'm tired of asking and nothing arrives. I ran out of bread and flour five weeks ago.
>
> I can make it with plantain or nothing. But the rest? ... Something to season; nor even once had I got onion, garlic, species, etc.
>
> Ever since the Colony exists people only feed on water, salt and fat. For my part I don't complain, but the rest?
>
> I don't have anything except corn, and the manioc and plantain will still take two months.[18]

If the 1912 long and mean winter meant that the road became virtually impassable in several parts, hence making the transport of provisions a true feat, the dry season that followed did not necessarily alleviate the situation in the colony. For instance, in February 1913 the Commander of the military division stationed in Puerto Asís, General Lucio Velasco, wrote to the President and the Minister of War notifying them of the worrying state of affairs there. The General reported that the low level of the Guineo River (tributary of the Putumayo) in Puerto Umbría had delayed for several days the transport of goods from Puerto Umbría to Puerto Asís, about 50 kilometres to the south, and noted that construction of the road between these two points had not yet started. In consequence, he complained that the soldiers had been 'very malnourished' and that if the drought were to continue, and he had heard this was highly likely, 'disaster will be imminent'.[19]

Despite General Velasco's dire omen, Puerto Asís survived its chaotic first year. In November 1913, the central government issued a law (Law 52 of 1913) that decreed the creation of a 'Junta de Inmigración' (Immigration board) in charge of promoting and supporting the colonization in the Caquetá and Putumayo, and for that purpose established an initial annual budget of 20,000 pesos. The Junta, chaired by the Governor of Nariño and the Apostolic Prefect, was assigned a wide array of tasks, ranging from the identification and allocation of baldíos[20] to the establishment and overseeing of the 'moral standards' to be fulfilled by current and potential colonos.[21] This last subject would be particularly observed, as evidenced by the strict 'Conditions of Morality' established by Las Corts in the Puerto Asís settler's guidelines,[22] or the policy set by the Junta, which detailed that in order to be eligible colonos should provide a 'certificate of morality' issued by their town's priest or mayor.[23]

With government's financial aid, administrative support provided by the Junta[24] and above all the obstinacy of the Catalan friars, Puerto Asís began to thrive. The Mission's annual reports shed some light on the colony's progress in the subsequent years. The 1914 report, for example, notes that there were 210 individuals living in the colony, without counting the army soldiers, and also that commissioners had been sent to Nariño to recruit another 500; as for the 'material progress', it mentions two sugar mills and a sawmill imported from the United States, abundant plantain crops – exceeding 50,000 plants – and a significant increase of pasturelands thanks to grass seed brought from Europe (*Misiones Católicas del Putumayo* 1914, pp.45–52). The 1916 report lists a hospital, 80 cattle belonging to the Mission, a big supplies shop for the settlers and four crews (totalling 90 men) employed on the Mission's enterprises, which included several plantations (coffee, cotton, sugar cane and rice, among others), construction works and cattle breeding (*Informe sobre las Misiones del Putumayo* 1916, pp.42–44).

By 1917, the number of 'white' settlers had reached 346 and there were 46 houses built, 250 cattle (owned both by the settlers and the Mission), and 720 hectares cultivated (*Informes sobre las misiones* 1917, pp.97–98, 118). Fray Gaspar de Pinell, currently in charge of the colony, enthusiastically wrote in June of that year that 'the houses are overcrowded and every day I find myself having problems accommodating newcomers. There are so many people that they consume two big steers each week' *(Informes sobre las misiones* 1917, p.77). A section of one of the plans of Puerto Asís (Figure 3.1) gives a sense of the colony's progress around that time, as indicated by the town's buildings and houses, which included a chapel, convent and orphanage (nos. 1 and 2), girls' school (no.3), army headquarters (no. 4), transport agency (no. 5), custom's house (no. 6), liquor agency and mail office (no. 7), slaughterhouse (no.8) and settler's houses (no. 10 onwards).

Two years later, however, this enthusiasm seemed to have melted away and the future of Puerto Asís looked bleak: the military contingent, which according the friars 'brought life and movement to the Colony', had been withdrawn; the government's financial aid had been reduced to half, and the friars complained that the budget was insufficient 'even to subsist', a situation that resulted in some families leaving. 'In consequence', the friars bemoaned the fact that the 'Colony, whose life is so crucial to support, has greatly diminished' (*Labor de los Misioneros* 1919, pp.33–34).

The missionaries did everything possible to overcome the most pressing problems of the colony. In order to deal with the issue of acclimatization, they brought a group of 'colonos morenos' (black settlers) from the province of Chocó, purportedly with very good results (*Informes sobre las misiones* 1917, pp.77–78). The task of 'reducing' the indigenous tribes

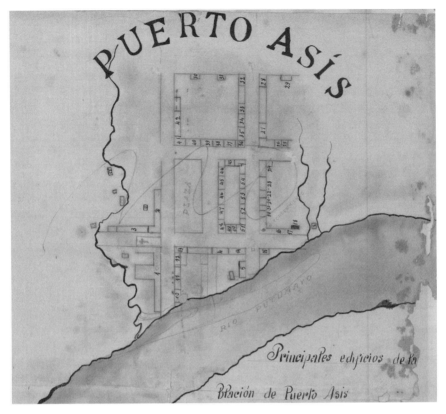

Figure 3.1 Section of plan of Puerto Asís, c. 1917.
Source: Archivo de la Diócesis de Sibundoy.

from neighbouring areas, mainly Kofán and Siona, proved more challenging. Nevertheless, the Mission partially solved this difficulty with the establishment of an orphanage managed by Franciscan nuns (Figure 3.2), which by 1916 already housed 100 indigenous children. The way to recruit pupils described in that year's report, which basically consisted of compelling their parents to leave them under the nuns' care, reveals that the term 'orphanage' was in practice a euphemism for the often harsh 'civilizing' methods deployed by the Capuchins (*Informes sobre las misiones* 1917, pp.86–87). The move in this case turned out to be doubly effective, since by retaining the children the missionaries not only facilitated the evangelization work amongst them, but also kept their families close. In addition to this, the Apostolic Prefect years later issued a stern regulation for the same indigenous communities, which required them, amongst others, to establish their houses and crops in a designated place near the colony, where they could be 'under the surveillance of the missionaries'.[25]

Figure 3.2 Puerto Asís orphanage, 1931.
Source: Archivo de la Diócesis de Sibundoy.

Despite the victories won against the hostile climate and the Putumayo natives, there was an obstacle the adept monks could never totally over-come. Having inspired the idea of Puerto Asís, the road from Pasto was expected to reach the colony in the same year as its foundation. However, it took more than two decades to be completed, and eventually became a major headache, keeping this frontier settlement stagnant for many years. Indeed, if the colony endured hard times at the outset due to the difficulty of getting provisions, this problem was aggravated when it came to trading its products. To deal with this problem, the Capuchins went as far as attempting to establish trade with the Brazilian city of Manaus, about 3,000 kilometres by river. In 1918, they persuaded the government of the importance of the project in both commercial and sovereignty terms, and eventually travelled to that city and chartered a 45-ton boat. The experiment was frustrated early on when the boat, packed with Brazilian goods and carrying Fray de Pinell, was arrested by Peruvian authorities and sent back to Manaus, allegedly for not having directed the ship to Iquitos on its way to Puerto Asís (*Labor de los Misioneros* 1919, pp.40–48).

This fiasco did not deter the friars whom, obsessed with the idea, soon embarked on an even more ambitious project. Again with the government's

Figure 3.3   Puerto Asís, panoramic view, 1928.
Source: Archivo de la Diócesis de Sibundoy.

support, in 1919 they imported a cargo ship from England to the Pacific port of Tumaco. From there, it was transported piece by piece on mules for nearly 500 kilometres across the Andes and the Amazon lowlands, and finally assembled in Puerto Asís (Figure 3.3). Then, and to the friars' dismay, they found that the boat's engine was too feeble to navigate the Putumayo River upstream (*Las Misiones Católicas* 1921, pp.94–98). This epic failure ended up burying intercontinental navigation projects for many decades.

As long as the government's concern regarding the foreign threat on the Putumayo lasted, the Apostolic Prefect's pleas remained effective, hence giving life, although artificial, to the colony. Yet, when awareness began to fade and support withdrawn, the vulnerable nature of the Mission's colonization project was fully exposed. Several Mission reports, printed articles, letters and telegrams sent to Bogotá recount the settlement's appalling situation.[26] Another significant event would have to occur, the armed conflict with Peru in 1932, for Puerto Asís to momentarily awaken from its languid existence. Meanwhile, it managed to survive, though in a critical state. The cry of its inhabitants in a telegram, one of many, sent to Bogotá in 1926 cannot be more telling:

> We *colono* compatriots form Colombians' outpost frontier with Peru, Ecuador, where love for Colombia sustains us despite all sorts privations, subjected by inhospitable forests and isolation from civilized region, conclusion road between Umbría and Puerto Asís needed … respectfully ask attention of your progressive spirit requesting deign to favour inhabitants Amazon basin breaking once and for all ten leagues wall virgin forest … trail to Puerto Umbría will be region's redemption, will facilitate entrance *colonos*, will end dangerous navigation causing too many victims, and will mark imperishable progress seal.[27]

Figure 3.4 Antonio Díaz carries on his back the jaguar or 'tigre mariposa'
hunted by him, 1930.
Source: Archivo Provincial de los Capuchinos de Cataluña.

'Privations', 'isolation', 'redemption', 'inhospitable forests', 'civilized'
and 'progress', are some of the words contained in the telegram quoted
above that form part of the dominant language present in the missionaries and settlers in the countless claims, grievances and appeals to the
central government. This language is also often expressed visually in the
Mission's photographs and maps, which convey both the missionaries'
material and spiritual 'achievements' and the frontier's inhospitable and
neglected condition (see Figure 3.4).
    The hegemonic nature of this vocabulary is revealed in how it
exhibits a shared grammar through which power relations are enacted
and expressed. In other words, as noted by William Roseberry,
hegemony is revealed here in how 'the words, images, symbols, forms,

organisations, institutions, and movements used by subordinate groups to talk about, understand, confront, accommodate themselves to, or resist their domination are shaped by the process of domination itself' (Roseberry 2004, p.361). In the specific history of Puerto Asís, this constitutes a language that powerfully reifies and naturalizes the antinomies between frontier and state, civilization vs. savagery, centre vs. periphery, progress vs. backwardness, and the relationship of inclusive exclusion that such antinomies sustain.

The following section tells the story of Sucre, another Mission settlement located in the Sibundoy Valley. This story is in many ways different to Puerto Asís, starting from the contrasting roles the road played in both cases. However, in another sense, this is a story that, like Puerto Asís, cannot be isolated from this vocabulary, whose different manifestations (literal, visual, cartographic) are crucial to understanding and further exploring such relationship.

Sucre

If the absence of the road was a lasting curse for the *colono* settlers of Puerto Asís, its arrival at the Sibundoy Valley entailed dramatic changes for its native inhabitants and the valley's physical landscape. The origins of this story, however, cannot be solely attributed to the road, and are largely related to a violent land conflict that, for several decades, confronted the Capuchin Mission, the Indians and the *colonos*.

Around the end of the nineteenth century, when the first 'whites' (mainly mestizo peasants from Nariño) began to occupy lands of the valley, which the natives claimed as their own, fierce disputes began to arise between them. The Capuchins, judging that conflicts could be avoided by segregating the spaces inhabited by both groups, and at the same time convinced that the presence of the new settlers in the valley would have a great 'civilizing effect' on the Indians, quickly came forward with a solution they deemed 'Solomonic': they persuaded the Sibundoy Indians to cede a land parcel in the eastern corner of the valley, and founded there the town of San Francisco in 1902, a settlement strictly conceived for 'white *colonos*'. One of the first descriptions of San Francisco comes from Miguel Triana, who in his 1906 expedition journal noted that the town was a tiny 'white colony' made of 30 to 40 small houses, and constituted 'the social symbol of the civilized race' (Triana 1950, p.359). Impressed by the beauty and fertility of the valley the engineer observed that its only inconvenience was the Sibundoy Indians, who represented an 'insurmountable obstacle for civilization and farming' (Triana 1950, p.361).

Despite Triana's judgement, the fact is that at the time he visited the valley, its population was predominantly indigenous, 2,700 against 270 whites according to the 1904 Intendancy Census, and the lands were legally constituted as *resguardo* (indigenous communal territories). This situation began to change with the coming of the road. As mentioned in the previous chapter, the road was not only conceived as a vital infrastructure for the missionaries' civilizing plans. Nariño's elites, attracted by the valley's lands, were eager to take advantage of the many opportunities the road would bring and actively supported the project. These expectations are not difficult to appreciate. The Sibundoy Valley, 50 kilometres east from Pasto, forms a plateau of approximately 8,500 hectares at an average altitude of 2,000 metres, whose rich soils, abundantly irrigated by rivers rising in the surrounding mountains and flowing through the valley, offer ideal conditions for agricultural and livestock production.[28] Yet, by the time the road was opened, the valley's lands were far from resembling the vacant and uncultured landscape portrayed by Triana or the Catalan missionaries themselves. For instance, the Inspector commissioned in 1912 by the government to evaluate the recently inaugurated road, observed in his report that the valley was 'heavily populated' and 'fairly cultivated' by the indigenous communities.[29] In another missive he also noted that although the Indians claimed ownership of the valley and resented the presence of 'white' settlers, there was 'great enthusiasm' on the part of the latter to obtain land grants there.[30]

The Inspector's remarks largely anticipated the drastic agrarian reform that was about to take place in the valley. The parts involved in this long and conflictive process, however, were not restricted to the Capuchins, the valley's natives, and a few greedy land-grabbers from Nariño. As some authors have noted (Brücher 1968; Chaves 1945; Wesche 1974), the early-twentieth-century colonization of the valley and the Putumayo foothills is largely associated with the prevalent land tenure system in Nariño, characterized by the excessive predominance of *minifundios* (smallholdings). The scarcity of land translated into the successive division of property into tiny and uneconomic plots, hence leading to an untenable situation marked by conflict and increasing poverty. This situation differed slightly from the general pattern in the neighbouring departments of Huila and Cauca, and broadly speaking the Andean region, characterized by the expansion of *latifundio* (latifundium) at the expense of indigenous and peasant lands.[31] The effect in both cases, nevertheless, would be the same and constitutes one of the most distinctive features of the country's twentieth-century agrarian history: thousands of landless people either forced to work on large landholdings under different, often coercive, labour systems, or fleeing to remote 'frontiers' attracted by promises of abundant free lands.

The history of Sucre can only be understood within this broader geographical and historical context. This history, on the other hand, began to materialize with the arrival of the road and more specifically with the sanctioning of a law (Law 51 of 1911) months before its inauguration. The Law's chief objective, as stated in its first article, was to facilitate the colonization along the road, and for this purpose decreed the founding of a town in the middle of the Sibundoy Valley, to be named Sucre.[32] Although it did not explicitly neglect the Indians' rights over the valley, the law stipulated that they had to prove those rights, whose defence would be assumed by a lawyer appointed by the government and paid by the department of Nariño. If their claims were rejected in court, the Indians would lose their current possession of communal lands, and instead be allotted two hectares of land per head in a designated place different from the one chosen for Sucre. As for the rest of the valley lands, they would be redistributed between the Mission, the future town and the department of Nariño, and the latter would be entitled to sell portions, in plots from 50 to 100 hectares, by public auction. Law 51 was amended by further laws (106 of 1913 and 69 of 1914),[33] which apart from defining the location, extension and other aspects related to Sucre, created a 'Junta de *baldíos*' (vacant lands board) in charge of putting the new regulations into practice and chaired by the Governor of Nariño, the Apostolic Prefect and five members of the Pasto City Council.

The indigenous claims, as the conflicts that began to unfold in the valley would soon demonstrate, proved to be fruitless. The story of how this happened is, however, less clear. There is evidence, provided by Friede, that confirms that at the time Law 51 was being passed, the Indians had in their possession title deeds, which they had referred to in a telegram to the President, proving their ownership of a considerable part of the valley's lands (Friede 1945, p.317). The document, known as the 'Testamento de Carlos Tamabioy', basically consisted of a will, dated from 1700, in which the Indian chief Don Carlos Tamabioy left his sons and his clan land he had purchased from the king of Spain.[34] Bonilla, the most severe critic of the Mission, argued that the Law was flawed since the payment of the lawyer assigned to defend the Indians was to be assumed by the department of Nariño, one of the parties involved in the process. Most gravely, he suggests, based on another person's testimony, that the missionaries intentionally mislaid or seized and hid the alluded to testament (Bonilla 1972, pp.108–109). Whether Bonilla's allegation is true or not, the fact remains that the Apostolic Prefect was determined to carry forward the reform. Certainly, the obstinate friar not only actively lobbied for the laws in question, but in a fervent pamphlet supporting the founding of Sucre he stated:

> The foundation of Sucre is imperative; and it will happen; because it must happen; it is needed by the poor of Nariño, it is needed by the Indians of Sibundoy Valley, it is needed by Pasto and above all it is needed by the Nation (*Informes sobre las misiones* 1917, p.28).

Once the laws were sanctioned, the missionaries rushed to put them into practice. The Indians, receiving the poorest deal in the new territorial order, angrily protested through letters and telegrams sent to Bogotá. One of these missives, a memorandum to the President signed by the Indians of Santiago – where Sucre was to be built (Figure 3.5) – shows their despair and indignation regarding the state of affairs. The beginning of the two-page letter reads:

> Tired of bearing a black existence in every sense on account of the Capuchin Mission, which has deprived us of our right of property over lands we opened up with much sweat and effort, we are today ceaselessly stripped of the best part of these lands by the alluded Mission, which doesn't respect any kind of authority and much less the laws.[35]

What followed was a denunciation of the burdens and punishments imposed by the friars, including the use of stocks, and a desperate plea

Figure 3.5   Santiago, 1920.
Source: Archivo de la Diócesis de Sibundoy.

to the President to take matters into his hands, for they declared: 'there has not been authority [in the Sibundoy Valley] to execute the mandates of the Supreme Government'; to which they added, contesting yet at the same time reinforcing the missionaries' rhetoric: 'it's been more than two hundred years since we stopped being *savages* and entered civilized life, and therefore we are not subjected to their authority'.[36]

Fray de Montclar dismissed the claims as mere stratagems instigated by the 'enemies' of the Mission, and particularly by large landowners from Nariño eager to grab the indigenous lands. According to the Prefect, the latter constantly tried to bring the Sibundoy Indians, who were 'minors barely entering the road to civilization', against the missionaries, so they could thwart the reform and seize their lands (*Informes sobre las misiones* 1917, p.34). Whether the claims were the sole initiative of the Indians or incited by the missionaries' foes is not what matters here. What matters is, first, that these claims were constitutive of rather than external to the hegemonic language of state and frontier and, second, that the 'excesses' of the Mission could hardly be avoided by the 'authority' of the 'Supreme Government', for the simple reason that the Mission embodied such authority and government. Thus, and despite the protests and fierce resistance of Indians and 'white' land-grabbers, the Apostolic Prefect's will prevailed. A Missionary chronicle filed in the Mission's archive of Sibundoy contains the following entry for 10th May 1916: 'Official foundation of Sucre (today Colón)[37] after tenacious opposition from whites and Indians which alleged rights over the lands reserved for the new town. The advocate of this foundation was F. Fidel de Montclar'.[38]

The 'reserved' lands for Sucre consisted of 10,000 hectares of terrain adjacent to the indigenous town of Santiago, and the place chosen for the new town was a spot seven kilometres east of the latter. Figure 3.6 shows a copy of the cadastral plan for Sucre sketched in 1917. Each polygon, marked with big numbers, corresponds to a separate chart of a larger scale (see Figure 3.7), which provides greater topographical and cadastral detail. The small grids, numbered from 1 to 418, represent the lots to be allocated or already allotted, most of which are ten hectares in area, as established by the Law 69 of 1914. The little square dots, meanwhile, indicate houses built outside the town, most of which are located along the road, fenced-in on most of its course as indicated by the dotted lines on both edges.

As with the Mission's statistical data, we can only take Sucre's cadastral plan as a partial, inaccurate and highly simplified representation of *reality*. As noted by James Scott on the relation between cadastral surveying and state-making, there is always a big gulf between 'facts on paper from facts on the ground' (Scott 1998, p.49). This gulf does not solely stem

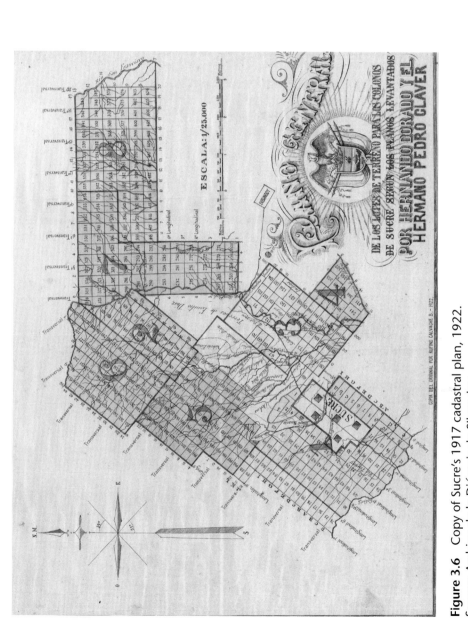

Figure 3.6 Copy of Sucre's 1917 cadastral plan, 1922.
Source: Archivo de la Diócesis de Sibundoy.

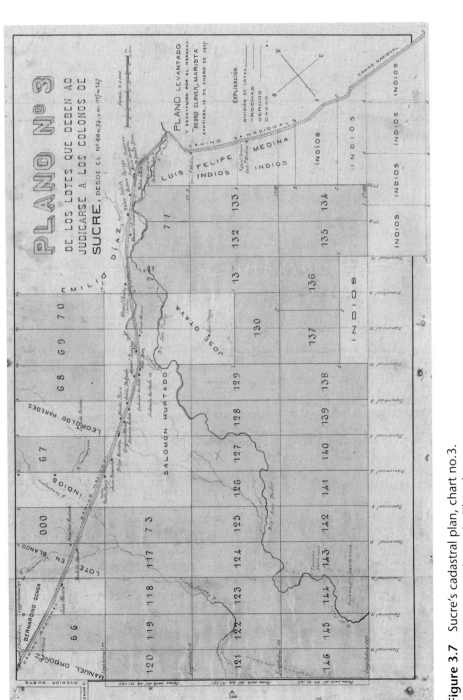

Figure 3.7 Sucre's cadastral plan, chart no.3.
Source: Archivo de la Diócesis de Sibundoy.

from the accuracy issues resulting from translation of data from field to paper, but from the fact that the legibility and order the surveyor aims to impose through the act of measuring and mapping is constantly altered and subverted at the level of practice. Nevertheless, the crucial significance of Sucre's plan lies not in the degree to which the cartographic projection resembles the physical and social space it aims to capture and transform, but in how through this distance the realm of representation not only appears detached from the reality it encloses, but is granted such power that the image is no longer a reflection of the object but the object a projection of the image.

This movement of estrangement and inversion that we find reflected in Sucre's cadastral plan, on the other hand, is hardly differentiable from the logic of state-making rituals discussed in the previous chapter. It is also closely connected to Reyes' South America's map discussed in the first chapter, though there is a crucial difference between Reyes' chart and Sucre's plan: whereas in the former we witness how the cartographer's rhetorical fiction acquires an illusion of reality, in the latter the distinction between fiction and reality and the binary conventions in which this distinction is sustained are no longer visible. And yet, it is precisely this absence that renders those conventions so overwhelmingly present in the image that we cannot avoid noticing them. Still, in order to bring them to the map's surface, we must consider Harley's remark on the nature maps, whose tracings, boundary lines, frames and grids, enclose as much as they exclude (Harley 2001, p.62). This is exactly what we are faced with when examining Sucre's plan: in filling the map with an intricate system of lines, dots and polygons, the cadastral surveyor simultaneously projects a new spatial and social order and silences the old one. Within this creative destructive process the natives, literally relegated to the margins of the map, are simultaneously included and excluded from this new order or, more exactly and going back to Agamben's topological structure of exception, 'included by means of exclusion' (Agamben 1998, p.7). As for the *official* character of this new order, it is emphasized by the country's coat of arms drawn on the bottom right corner of the map, and whose motto reads 'Freedom and order'.

***

According to Vilanova (1947, vol.2), the first settler arrived in Sucre on 2nd February 1916. By the end of that year, Fray de Montclar reported that 250 families had been already assigned home sites and land plots. The friar was optimistic about the increasing flow of *colonos* and praised the advantages of Sucre for the Indians in the following terms: 'As we have said before, in order to become civilized, the Indians require the contact of the whites; through them, they practically learn their customs which, as bad as they can be, belong to civilized peoples, and are thus less repugnant' (*Informes sobre las misiones* 1916, p.26).

The 1917 Mission's report included a detailed census of Sucre, which shows its rapid growth, significantly faster than Puerto Asís, and particularly remarkable if we consider that the town had been in existence for barely a year. The census, carried out on 23rd June, indicates that at that date, 68 houses had been built and 80 were under construction, the number of settlers had reached 835 and the hectares farmed 225; however, a note at the bottom of the statistical appendix states that a month later (18th August), the number of houses built had increased to 95 and those under construction to 93 (*Informes sobre las misiones* 1917, pp.102–114). According to the 1919 report, the number of settlers had increased to 2,000 and the houses built to 200. The town, adds the report, already has a church, a school and a cemetery (*Labor de los Misioneros* 1919, p.39, 101).

The 1917 statistical appendix also counts 63 head of cattle owned by settlers from Sucre, and 912 in the whole valley, out of which 512 belonged to settlers and 400 to the Capuchin Mission (Figure 3.8). This detail is important since cattle, as much as humans, played a vital role in the twentieth-century political ecology and environmental history of the Sibundoy Valley. The Mission's statistical reports for the following years

Figure 3.8 Capuchin cattle farm in the Sibundoy Valley, c. 1912.
Source: Archivo Provincial de los Capuchinos de Cataluña.

omit information on this particular subject, although there exists evidence of the rapid expansion of pastureland and livestock in the Apostolic Prefecture, particularly the Sibundoy Valley. In 1916, Fray de Montclar reported that up to that year more than 500 rolls of barbed wire (500,000 metres) had been used to fence the Mission's cattle and agricultural lands in Sibundoy and Puerto Asís (*Informes sobre las misiones* 1916, p.43)

Four years later, in August 1920, the Capuchin friar submitted a thorough report to the Ministry of Public Works about the road and some of the most significant changes brought by it to the region. The road, noted Montclar, had been particularly beneficial for the development of agriculture and cattle ranching, to the point that this activity was 'transforming the region and its towns completely'. 'Ten years ago', he stated, 'there were just a few heads of cattle in the whole Territory, not a horse, and nowadays there are various thousands, and a regular number being exported each year'.[39] In 1922, he estimated that just in the valley the number of bovine cattle was no less than 7,000;[40] and in 1954, according to data provided by Bonilla, this figure had reached 14,000, 'almost all of which belonged to the white population' (Bonilla 1972, p.225). These numbers alone shed some light on a long and complex story that is yet to be written, and which would help understand the conflictive society–nature relationship shaping the landscape of the Sibundoy Valley throughout the twentieth century.

Even by the missionaries' standards, Sucre was far from a successful experiment. In 1922, the Apostolic Prefect wrote a short account of Sucre in which, although noting that the town 'prospers day by day', he also acknowledged that that same progress had been the cause of the persistent usurpation of the lands allotted to the Indians following the Law 51. The Capuchin superior washed his hands of this grave issue arguing that despite the Mission continually urging the Indians to preserve their lands, 'unscrupulous' persons often 'induced' them to sell. In addition, he stressed that those purchases were illegal since under the national laws, the Law 89 of 1890 specifically, the *resguardos* were inalienable. The friar thus concluded that unless the Indians were prevented from selling their plots they would soon be 'without an inch of land and reduced to slave status'.[41]

Montclar's argument could not conceal a central contradiction: in having crusaded for the valley's agrarian reform, the Mission deliberately denied the indigenous property right claims and instead opted for the allocation of individual land parcels, hence breaking the communal possession of lands guaranteed under the legal figure of *resguardo*. The long-term effects of the Capuchin 'reform', just considering the issue of land tenure, were dramatic. Rolf Wesche, a German geographer who studied the colonization of the region during the early 1960s, observed that of the approximately 2,500 families then occupying the valley, about 300 were landless, 700 owned land plots of less than three hectares and 300 from three to five hectares; most of these families, which according to Wesche 'eked out a miserable livelihood', belonged to the indigenous towns of Santiago and Sibundoy (Wesche 1974, pp.44–49). The Mission, on the other side of the spectrum, was the largest land owner in the valley, its properties amounting to more than 1,000 hectares. This situation, which brought growing criticism upon the Catalan friars, would eventually explode with the publication of Bonilla's book in 1968 and finally led to the departure of the Mission from the Putumayo two years later (Bonilla 1972, pp.274–276).

This is not the place to discuss in depth the Sibundoy agrarian conflicts or the role played by the Mission in this story. As mentioned earlier, in highlighting some aspects related to the history of Sucre and Puerto Asís, the aim of this section has been primarily to situate the road within the spatial history of the frontier. The road, as emphasized, not only embodied a powerful hegemonic language that we can trace back to characters such as Codazzi and Reyes, but became a paramount symbol of the rituals and practices of state- and frontier-making. In this sense, Fray de Montclar was certainly right when he proclaimed the road as being the 'precursor' of radical transformations in the territory governed by him. The colonization of 'vacant' lands by both 'white settlers' and foreign animal and plant species, the extraction of all kinds of products from the forest and the dispossession of the natives' lands and labour, are just some of the interwoven dynamics the road actively assisted. Yet, as most of the projects conceived by Fray Fidel de Montclar and his relentless army of Catalan friars, the road hardly matches the triumphal image it projected in its inauguration rituals and celebratory accounts. In this sense, as we will see in the next section, the road was a project that, more than any other, speaks not just of the power and effects of myth and ritual but also of the large gap between myth, ritual and everyday reality.

The annihilation of theory in practice

In December 1911, just a few days after Estanislao de Las Corts had announced the imminent arrival of the road to Mocoa, Fray de Montclar sent a letter to the Minister of Public Works. The Apostolic Prefect, with the project of Puerto Asís well underway and well aware of the government's desperate need of a way to access a navigable point on the Putumayo River, did not limit himself to announcing the good news. Instead, he warned the Minister of the vital importance of extending the road to the future colony and, in order to persuade the officer, argued that having already overcome the 'Andes barrier', the missing section was an easy enterprise. Keeping the same number of workers (1,200) and with an additional budget of 30,000 pesos, the Mission, he assured, would be able to conclude the road in three months.[42]

Three months later, and after having sent another missive to the Minister with some pictures attached of the Mission's heroic deed ('what remains is extremely easy, there are no more cliffs to break' he emphasized this time), the insistent friar had his way: the Ministry committed to continue sending 10,000 pesos monthly and to evaluate the Apostolic Prefect's request of raising the budget so as to increase the number of workers to 2,000.[43] The road, nevertheless, and despite Montclar's confident and ambitious prediction, did not arrive at Puerto Asís in 1912. When it eventually did, 19 years later, in 1931, the Apostolic Prefect would not be there to witness the event (Figure 3.9).

There is little doubt that Fray de Montclar highly underestimated the difficulty of building the almost 90 kilometres of road from Mocoa to Puerto Asís, and the friar himself would later acknowledge that this undertaking proved far from 'extremely easy'. But the factors explaining

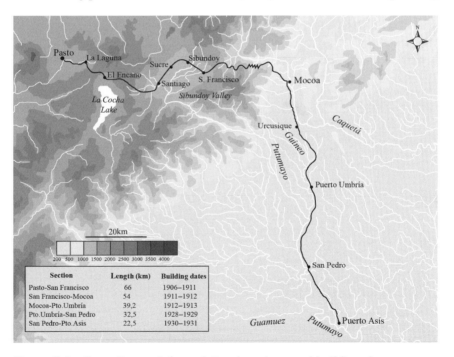

Figure 3.9   Pasto-Puerto Asís road. Road sections and building dates, 1906–1931.
Source: Elaborated by author. Based on AGN, MOP, Vol.408, n.f.; Vol.3273, fol.241.

why the project took that long and was so difficult to complete are many and have little to do with Fray de Montclar's miscalculation. Taken together, however, those factors expose the gap between the coherent discursive logic of the state and the conflictive, contingent and often messy everyday state practices.

A first aspect that constituted a perpetual obstacle to the road was bureaucracy. The 1886 Constitution had replaced the prior federalist system with a centralist regime, a shift partly aimed at strengthening the political weight of Bogotá over the country's states (henceforth called departments). Although in theory this change did not affect the *Territorios nacionales* (chapter 2), in practice it made things more complicated. The Putumayo, for instance, having once belonged to the large *Territorio del Caquetá,* itself part of the former State of Cauca, was named an Intendancy in 1905 and a Special Commissionership in 1912, administrative units heavily dependent on Bogotá. However, for many administrative, economic and political matters, it also relied on the department of Nariño, a situation that in some ways persists today.

The road did not escape this double dependency on Bogotá and Pasto. As noted, the central government delegated the direction of the road works to Fray de Montclar but at the same time left the Governor of Nariño in charge of managing and overseeing the funds. For the Mission, this basically meant that it had to submit regular reports both to the Governor in Pasto and the Minister of Public Works in Bogotá, a situation that not only placed a heavy burden on the friars but often translated into bureaucratic inertia and delays in disbursements. This problem, moreover, was further magnified if we consider the long response times of official correspondence, an issue related to the long distances between the Putumayo and Bogotá and the country's precarious transport network. Therefore, while a telegram sent from Mocoa or Sibundoy to Bogotá usually took about a week to get a response, letters and reports could normally take between 30 and 50 days.[44]

A typical case exemplifying the above occurred in March 1912, around the same time the inauguration of the road to Mocoa was taking place. Just a few days before travelling to Mocoa with the commission appointed to inaugurate the road, Fray de Montclar was stuck in Pasto desperately trying to obtain a disbursement from Bogotá he had been waiting for since early February. The money, as the friar explained in a letter to Fray de Cardona – in charge of the road to Puerto Asís – had been frozen in Tumaco's customs house for several weeks. In order to deal with the anxious workers, the Apostolic Prefect persuaded them to continue working for no pay while he resolved the impasse, even using the threat of no longer offering them work. Meanwhile, he had written to Bogotá in the hope that they could sort the problem from there, and blamed the incident on the customs officer's incompetence.[45] A week later, on 8th March, he finally received funds from Tumaco, although just half of the expected sum. He therefore sent another telegram to the Minister of Public Works reporting the problem and complaining about the recurrent delays in the payments.[46] This time he got a prompt reply saying that the whole problem was due to a change of customs officer but that it had now been solved for good.[47]

The sporadic complaints about payment delays show that this situation not only persisted but sometimes resulted in workers' insubordination and desertion.[48] Bureaucratic difficulties, moreover, were far from restricted to delays in disbursements, and included budgetary constraints, slow decision-making processes and temporal suspensions of contracts.

If bureaucracy constituted a constant hassle for the Mission, everyday problems with workers and road users were never absent. Amongst these, the one that caused most damage to the road and became a perpetual headache to the friars was the dragging of wood along the road. This practice deserves special attention as it illustrates how the road was appropriated in ways different from those originally conceived, and how the coercive measures to control its users and uses were constantly resisted.

The extraction and transport of wood and charcoal constituted a traditional livelihood among the indigenous peoples inhabiting the surrounding areas of La Cocha Lake, 20 kilometres east of Pasto. This activity was highly damaging to the road, especially during the rainy seasons, when the heavy logs dragged by oxen along its surface formed deep trenches and quagmires. As early as 1909, when the section of the road from Pasto to La Cocha had just been concluded, Las Corts notified the Governor of Pasto that due to this activity, several stretches were impassable and had now to be rebuilt. At the friar's request, the Governor issued a resolution later that year ordering the placing of permanent guards along the road to prevent this trade, a measure that soon proved futile. Six months later, in June 1910, the guards had all resigned arguing that they did not have the support of the Nariño authorities and were a constant 'object of mockery'.[49] In October Fray de Las Corts, tired of sending complaints to Pasto with no results, opted to press the Governor by notifying the Minister of Public Works about the time and high sums of money wasted repairing the road's damaged stretches.[50] This time the priest's pleas apparently had some effect. The Governor sanctioned a new decree contemplating several police measures for road users including fines for the dragging of wood,[51] and even went as far as requesting, although unsuccessfully, that the army take control of the situation.[52]

In July 1912 Rufino Gutiérrez, a fiscal inspector sent by the central government to evaluate the road works, noted that there were currently some guards posted along the road, which had been given the order to imprison the Indians using beasts of burden, even if unloaded. The natives angrily protested against the measure, arguing that it was especially unfair not only because it only applied to them but also because they had worked on the road for no pay. The prohibition, as the same officer acknowledged, was certainly unjust, particularly considering that the Capuchins had built a section of the road over the old trail the Indians had opened to transport their goods to Pasto. Gutiérrez therefore persuaded the Governor of Nariño to lift the ban and allow the traffic of animals as long as they didn't drag the loads, a condition that according to the officer the Indians 'happily accepted' (Gutiérrez 1921, p.339). Yet this 'truce' didn't last long, and a few weeks later the situation was back to the beginning.[53]

Despite endless claims, coercive measures and conflicts, the Mission and civil authorities could never put an end to this practice. Ten years later, Fray Canet del Mar, at the time in charge of the road, sent a discouraging report to Bogotá in which he listed this activity, together with rain, as being the most critical problems currently affecting the Pasto-Sibundoy section. 'The dragging of wood', lamented the Capuchin, 'is one of the abuses we have tried to avoid most and yet without success, neither through the surveillance of police nor through the Governor of Pasto, who has intervened during different times at our request'. To convince the Minister of the magnitude of the problem, the friar described the part of the road in the worst condition as a 'mass slaughterhouse', and assured that if it remained unrepaired the Putumayo would soon be left 'completely cut off from the rest of the Republic'.[54] The high intensity of this traffic, as revealed by a census of products transported on the road in that year, suggests that despite his fatalistic tones Fray del Mar's gloomy picture may not have been far from reality. According to the census, between March 1922 and April 1923, 20,650 timber pieces of different dimensions and 62,400 arrobas of charcoal (around 390 tons) were transported from the Putumayo to Pasto.[55]

Once again, the missionaries attempted to curtail this practice, on this occasion through the Putumayo's civil Commissary.[56] This time, the Indians sent the following telegram to the President where they vehemently expressed their frustration about the prohibition:

> Putumayo Commissary, in agreement Apostolic Prefect, under pretext preserving the east road, have forbidden the use of such road with oxen, hence they preclude us to exploit our lands ... We built the road not long ago and maintained it with our own hands. Can it be possible, equitable, advantages civilization only favouring privileged classes, Indians continue to be serfs? Respectfully entreat stop measure regressive Colony.[57]

This appeal, starkly exposing both the exclusionary politics of the road and the indigenous subversion of the 'whites' colonialist rhetoric, apparently remained unanswered. Thus the Indians, powerless to assert their rights within the realm of law, continued to negotiate their right to the road through the messy terrain of everyday conflict. At this point, however, the missionaries were exhausted from fighting against an enemy much stronger and more harmful to the road than the Indians' oxen.

This enemy was rain. Rain heavily damaged the road year after year, reminding everyone how fragile had been the powerful image of the Capuchin's *magnum opus* as man's exultant conquest of nature. The struggle against the rain, although varying in intensity, was never really absent from the history of the road, from its opening to its decline in the late 1920s. Still, of all the battles fought during those years, the one which took place in 1912 was exceptional since there was no other occasion when the line between loud triumph and failure was so dramatically exposed and staged.

The rainy season in the Putumayo usually extends from May to October and is characterized by abundant rainfall, especially in the *piedemonte*, a phenomenon associated with the dense clouds formed by the collision of trade winds with the Andean mountains. During this period, the monthly precipitation levels normally reach 400 millimetres and the flow of the rivers descending the *cordillera* often becomes torrential. A technical report of the road elaborated in 1913 gives a sense of the difficult obstacle water represented for the construction of the road: between Pasto and the Sibundoy Valley, the road had 50 culverts and pontoon bridges of different widths; from there to Mocoa, 72 culverts and small bridges from four to eight metres long, and 12 between ten and 12 metres; and from Mocoa to Puerto Asís, 71 culverts and bridges from three to six metres and 18 'large bridges' of lengths or spans varying from 12 to 30 metres (Figures 3.10 and 3.11).[58]

Every rainy season, the infrastructure suffered serious damage. However, 1912 was a very critical year for the Capuchins, not only because the rain hit harder than usual but since no sooner had they completed the road from Pasto to Mocoa, than they embarked on construction of the missing section to Puerto Asís. As the newly inaugurated section demanded regular maintenance and consolidation works in several parts, the friars now had to divide efforts between both sections, a burden that proved too much when the rains arrived.

It is difficult to establish when exactly it started raining in 1912, but in June, just three months after the road to Mocoa had been completed, the situation was critical. On the 21st June, the Apostolic Prefect sent an anxious note to Fray de Cardona urging the friar to meet him in Mocoa

Figure 3.10 Culvert (bottom right) and bridge (centre left) on the Mocoa-Urcusique section, 1914.
Source: AGN, MOP, Vol.1410, fol.66.

Figure 3.11 Fray Florentino de Barcelona (on horse) crossing a suspension bridge over Pepino River, Mocoa-Urscusique section, n.d.
Source: Archivo de la Diócesis de Sibundoy.

at once to discuss the likely suspension of the works. 'The carriers say that', reads the note, 'it is no longer possible to take a step forward since the *páramo* [Bordoncillo] is completely ruined with large depths, as is the case of Portachuelo and some parts around here. We are going to be cut off and unable to bring provisions'.[59]

A few days later, the recently appointed Commissary, General Joaquín Escandón, wrote to the Minister of Government confirming the gravity of the situation:

> Regarding to state of the road from Pasto to this place [Mocoa] I am afraid to inform your honour that thanks to a strong and prolonged winter, most of the road has suffered grave damages due to collapsing of slopes and platforms and destruction of palisades in the flat and swampy areas, to the point that the traffic, even for pedestrian travellers, is now extremely difficult. In consequence, the missionaries decided to remove the crews working in the section from here to Puerto Asís in order to employ them in recomposing the road to Pasto.[60]

As rain continued to take its toll on the road, criticisms of the missionaries, not so long ago loudly proclaimed as 'heroes of civilization', began to pour in. One of the Mission's most bitter enemies took advantage of the calamitous situation and sent an acrid missive to the Minister of Public Works sentencing that 'the government has made the biggest mistake in trusting the Spanish Capuchins, ignorant of engineering, with the opening of the most important road for the Republic'; to which he next implored the Minister to 'appoint an able body of engineers, honest, scientific; and put them in charge of the road and the military colonies'.[61]

The Minister, possibly fearing that the negative campaign against the Mission could eventually involve him, and aware that the whole road was too great a burden for the friars, chose to hand over the Pasto-Mocoa section to a civil engineer from Pasto, a measure that Fray de Montclar took as a personal offence.[62] Visibly irritated and anxiously trying to persuade the Minister to revoke the measure, he wrote to him: 'I foresee very serious problems as a consequence of the [new] arrangements. I think we don't deserve such a snub'.[63] In fact, what Montclar feared most was that as a consequence of losing control over the first section of the road, the one under construction could be left adrift.

The Apostolic Prefect's concerns, as it soon turned out, were not unfounded. In September, as the rains intensified and the contract to meet the urgent repairs in the Pasto-Mocoa section awaited the bureaucratic procedures in Bogotá, the crews working on the Mocoa-Puerto Asís section and the colony itself were on the verge of collapse. Fray de Las

Corts didn't cease sending dramatic updates of the situation to the Apostolic Prefect. 'Since yesterday night until now, 4:00 p.m., the rain has not ceased for an instant and the rivers are flowing over the mountains' wrote the friar from Mocoa on 1st September, adding that in consequence he had to postpone his departure to Puerto Asís.[64] On 17th September, still stuck in Mocoa, he despairingly noted:

> The road is out of provisions. The workers do not work because they have nothing to eat, many have left from hunger … My absence from Puerto Asís had prevented me from sending them corn, I have given them some but it's not enough. Most of the Colony's workers have left from fear and there is no one to harvest [the] corn.[65]

In October, Julio Thomas, the new engineer in charge of the Pasto-Mocoa section, wrote a meticulous report describing the current state of the road and the most urgent repairs and improvements. Thomas acknowledged the Mission's work, especially in the difficult stretch from San Francisco to Mocoa, which he judged as 'the main achievement of the Capuchin priests'. He referred to the architect of this section, Fray de Las Corts, as 'a vigorous man who defeated the abrupts [sic] at all cost', and yet added: 'perhaps without considering the technical aspects'. The engineer pointed out serious flaws in the road design and construction employing a technical jargon that made the missionaries look like naive amateurs. This clinical diagnosis regarded the organization of works as 'extremely defective', the solidification materials as 'inconvenient', the bridges as 'poorly built' and the road width as 'inadequate for traffic needs'. Even so, and perhaps conscious that earning Fray de Montclar's enmity was not a good idea after all, he attributed most of the problems to the government's pressure on the friars to get the road done in the shortest time possible.[66]

The Apostolic Prefect, not unexpectedly, not only did not take Thomas's observations well, but counter-attacked immediately. In early November, while an official Commission sent upon his request to evaluate the Mission's work visited the road, he telegrammed the President saying: 'National interest compels me to come to Y.E. [Your Excellency]. Road maintenance neglected … Procedure Engineer indicates interests impeding access Putumayo'.[67] The commission's report was released on 14th November, and for the most part was favourable for the Mission. The authors, two of which had been appointed by the Governor of Nariño and one by the Apostolic Prefect, praised the friars for the economy and speed of the works, attributed the current damages to the harsh rains and mostly to the pending consolidation works, and suggested some rectifications to the original layout and solidification works on most of the road. In summary, they noted that of the 150 kilometres built from Pasto to Puerto Asís only 15 kilometres could be considered to be 'in bad condition and dangerous'. The report also included a section on 'observations' written by Fray de Montclar, where the priest blamed the current problems on the government's rush, the 'extraordinary and never before seen winter', and the very high traffic of pack animals. Most remarkably, and in a statement that plainly exposed the fragile nature of the state myth so powerfully embodied by the road and reified by its inaugural rituals, the Mission's superior posed the following question: 'Will it be claimed perhaps that, because the missionaries built this road, it is not subjected to the *laws of nature*?'[68]

The 1912 winter certainly seems to have literally dragged the Apostolic Prefect backward to a Hobbesian state of nature. As it became evident that rain was an adversary he could not control, he lashed out against the critics of his prized project, often labelling them as 'enemies', a situation exposing the friar's fear of anyone contesting the Mission's authority. The dispute with Thomas concerning the Pasto-Mocoa section was just beginning and would gradually escalate into a frontal war of mutual accusations, gossip and intrigues through telegrams, letters and the press.

The Mocoa-Puerto Asís stretch was no better and would soon become another battlefront. In December the Civil Commissary, General Joaquín Escandón, the same who had written the laudatory report of the inauguration event, telegrammed to Bogotá notifying that the delay in the construction of the last 50 kilometres of road meant that this part had to be travelled by river, causing several shipwrecks and human losses. Escandón explicitly made the missionaries responsible for the accidents, noting that they had 'resisted' widening the existing trail to allow overland traffic.[69] This telegram would mark the beginning of a harsh conflict that soon spread from the road to the ongoing land disputes in the Sibundoy Valley, and in which the Apostolic Prefect would eventually prevail over the civil officer.

The quarrels and disputes in which the Apostolic Prefect engaged ultimately led to a vicious circle with no end in sight. In conceiving the road as a vital appendage of the Mission's dominance in the Putumayo, the Capuchin priest fiercely reacted to any criticism or attempt to snatch it from his hands, consistently labelling the road's critics as 'enemies' of the Mission, a situation exposing the friar's fear of anyone contesting such dominance. This in turn led to endless conflicts where the most affected part was the road itself. Nowhere was this deadlock better portrayed

than in the fragment quoted below from a letter sent by Fray de Las
Corts to Fray de Cardona in January 1913:

> I tell you with deep regret that the current situation of the road is I believe
> the most distressing. Numerous times I told the F. Prefect to leave the
> [road] works; he had done so gladly but the situation is so terrible in every
> sense because of the incessant wars of enemies and friends, that there is no
> way to know where to steer the course in the midst of such a furious
> squall. God help us![70]

The 'squall' did reach its height a few weeks later, although in a totally
unanticipated way. On 14th February, Fray de Montclar received a telegram from the Minister of Public Works ordering the indefinite
suspension of the works due to lack of government funds. To the Mission,
owing a large sum of money in back wages and provisions and having
completed just one-third of the Mocoa-Puerto Asís section, the measure
came as a shock. The Apostolic Prefect, having declared a year earlier
that this section was an 'extremely easy' task and could be accomplished
in three months, had no choice but to admit that his estimate had been
far too premature. However, in a public letter published in Pasto and
entitled 'The suspension of the road to Putumayo and its causes', he
fervently attributed the Mission's 'lack of caution' and 'financial indis-
cretions' to its 'excessive love' for the two 'great ideals' it had been
entrusted with: the 'defence of Colombia's sovereignty' and the 'Christian
civilization of the Putumayo'. Then, as the skilful statesman he was, the
Capuchin sovereign wrote: 'If all this is a crime, we plead guilty before
the nation and the Department [of Nariño]'.[71] Fray de Montclar must
have thought that thanks to his infallible argument he would get away
with it, and it certainly represented a victory for the Mission. Yet, what
the friar perhaps did not realize was that the drama of the road, far from
over, was just beginning.

Stagnation and decay

The years that followed the 1912 events can be considered to a large
extent to be a cyclical repetition of the same drama, although performed
with less intensity and by occasional novel actors on stage. Fray de
Montclar's rhetoric defence following the suspension of the road
managed to persuade the government to pay all the debts the Mission
had contracted, which in February 1913 amounted to 26,500 pesos.
However, the road was eventually removed from the hands of the
Mission. In June of that year, the section under construction was left in

charge of the military force stationed in Puerto Asís, and in March 1914 Thomas was appointed Engineer Director of the entire road.[72] The new arrangements did not deter the Apostolic Prefect from continuing to conspire against Thomas by every possible means, from sending telegrams to Bogotá and publishing press articles accusing the engineer of having abandoned the works and deceiving the government, to encouraging insubordination among the workers and foremen. Thomas initially replied in the same terms, but in the long run, the Mission ended up being too big an adversary, and the engineer finally resigned from his post in 1914.[73]

In August 1915 the government launched a public tender, and the winning contractor was committed to finish the missing section and to repair and consolidate the built sections within 26 months.[74] A year-and-a-half later, Fray de Montclar, exasperated by the little progress made so far, began a new campaign against the contractor. He sent a memorandum to the Minister of Public Works regretting the total 'incompetence' and 'indifference' of the contractor towards the road and urged him to rescind the contract at once, assuring that with the current procedures the road would not be completed 'not even in twenty centuries'.[75]

In August 1917 and following the friar's denunciations, the Minister sent an inspector to evaluate the state of the road, who concluded that the contractor had not met his obligations and that the road was totally neglected.[76] The government finally cancelled the contract in December,[77] an occasion that seemed propitious for the Apostolic Prefect to regain control of the road. Therefore, after the same inspector recommended handing the road back to the Capuchins, whom he described as 'very laborious, selfless, and competent',[78] the friar, noticeably triumphant, wrote to the Minister in May 1918 saying that he had 'no wish' to take charge of the road, and yet he added that: 'but the deep certainty that we have, confirmed by the facts, that the road will not be finished unless the Mission takes care of it, prompts us to face every difficulty that might arise and to ignore the calumnies and intrigues of our enemies'.[79]

The road, then, returned once again to the Capuchins' hands, though not as soon as Fray de Montclar had wished: budgetary constraints, bureaucratic obstacles and inertia in Bogotá and Pasto delayed the contract with the Mission for 15 months.[80] Meanwhile the road, neglected and left to the mercy of nature, continued its decay. This desolate landscape was condensed in a short telegram sent from Mocoa to Bogotá in June 1919: 'National road totally abandoned. From here to San Francisco thirty landslides; from here to Umbría, all bridges destroyed. Winter continues harsh'.[81]

The Mission took charge of the road in August 1919 and during the next four and a half years struggled against its customary human and non-human foes. Then, in May 1924, the Ministry of Public Works suddenly issued a resolution declaring the contract with the Mission to be dissolved due to 'unsatisfactory execution of works'.[82] The Apostolic Prefect, enraged, fiercely refuted the government's verdict in the press arguing that the meagre budget assigned to the road was barely enough to keep the built part passable.[83] Regardless of who was right or wrong, the fact is that since 1913, when the government withdrew the road from the Capuchins for the first time, it had not advanced a single kilometre beyond Puerto Umbría, half way between Mocoa and Puerto Asís. A new contract was signed in June with another engineer from Pasto and Fray de Montclar was designated as 'Inspector of Works'. This arrangement, however, lasted scarcely six months. In February 1925, the government issued another decree establishing that due to the small allowance voted in Congress for the road that year, the position of engineer was suppressed and the road would be left in charge of the Apostolic Prefect, now titled 'Manager Inspector'.[84]

The new monthly budget assigned to the road, 416 pesos in contrast to the peak of 10,000 reached in 1911, was certainly a derisory sum for maintaining the existing road and completing the missing 50 kilometres from Umbría to Puerto Asís. Although another 30 kilometres of road were built during the next four years, its critical situation hardly improved. During that time the central government, despite the incessant claims of the missionaries and settlers, suspended or delayed the payments persistently arguing fiscal constraints and budgetary crises.[85] This situation reached a dramatic climax in 1929, when no budget was approved for the road. In December 1928, as soon as the news reached the Putumayo, Fray Canet del Mar, currently in charge of the road, unsuccessfully wrote to Bogotá imploring the measure to be reconsidered.[86] During the subsequent months, the Capuchin friar and the same settlers who years before had come to the Putumayo driven by the illusion of the road, sent numerous letters and telegrams to Bogotá begging for a solution and reproaching the 'abandonment' and 'neglect' to which they were subjected by the government. The following telegram, authored by a group of inhabitants from Mocoa, bluntly summarizes the frustration and despair of the frontier *colonos*:

> If government left us complete oblivion, we will die isolated … widespread discouragement discontent *colonos*, we have right to be assisted preferentially rest of nation because we have diminished at the expense sacrifices sustaining unscathed sovereignty home soil; we don't ask for roads, railroads but demand conservation only gateway. We hope to be

assisted, otherwise will emigrate forfeiting what has costs us plenty of sweat, with prejudices government's good name, which neglects weak peoples, ignoring we all have right to budget because contribute with contingent and our own lives in unhealthy climates afar prerogatives civilized world.[87]

No response is recorded for this message. Others were eventually answered, although replies are almost invariably the same: a four- or five-line impersonal missive regretting the situation and yet informing that problems of 'fiscal order' precluded any short-term solution.[88]

As happened two decades earlier, the road would have to wait for another exceptional event to momentarily escape its neglected condition. This moment came in 1932, when the long-feared military conflict with Peru finally occurred. On 26th September, less than a month after conflict started, the Pasto-Puerto Asís road, the last 22 kilometres of which had been completed in 1931, was classified as a 'national defence' road. Its repair and improvement were graded as of 'immediate urgency' and for this purpose a budget of 120,000 pesos was allocated.[89] Three months later Rafael Agudelo, the new engineer in charge of the works, submitted an extensive report where he strongly argued that apart from the urgent widening of the whole road to allow vehicle traffic, the section between San Francisco and Mocoa had to be replaced totally. The engineer's verdict, marking both the death sentence to the Capuchins' most praised section of the road and the beginning of another never-ending story, reads:

> This road, as it is now, had existed for many years and, due to its bad technical conditions, none of its ends had been achieved, neither the cheapening of transport nor the development of public wealth. Neither can it be concealed that, as a strategic way, it is little less than an eyesore.[90]

Fray de Montclar was no longer there to oppose the engineer's ruthless judgement. Sick and aged, the once almighty Apostolic Prefect had left the Putumayo for good in 1928. He died in Spain six years later, on 21st March 1934, at the age of 66. Various obituaries highlighted the road as his most significant mark, and recalled one more time the story of how thanks to the Catalan priest this 'utopian' ideal had become a reality.[91] A chronicle narrating his posthumous tributes celebrated across the territory of the Prefecture effusively declared that:

> After his gigantic and uneven struggles against a rebellious nature, against savages obstinate in fatal superstitions, against countless emissaries of darkness, his funerals celebrated in the Mission of Caquetá constituted

a new triumph. There, a transformed nature could be observed. Mourning his death and praying for the soul of his benefactor, attended once savage tribes that, in ignorant and tragic opposition, had confronted all the works of progress that would come to redeem them.[92]

State and frontier revisited

Fray de Montclar's posthumous triumphal image strikingly contrasts with the countless reports, telegrams and letters regarding the deplorable state of the Mission's flagship project during his last years in the Putumayo and Spain. Figure 3.13, showing the Apostolic Prefect supervising the road works in 1927, 15 years after the road was supposed to have reached Puerto Asís and not long before he moved to Spain permanently, makes it hard to believe that at this point the friar remained as enthusiastic about this decaying infrastructure as he was in 1912. The many obstacles against which he battled incessantly, for the most part in vain, surely shook his faith in this 'redeeming enterprise'. In the end, nevertheless, it was not bureaucracy, or the dragging of wood, or even rain that ultimately frustrated his dream of seeing the road concluded. Those constituted everyday obstacles that largely account for the turbulent history of the road, and lay bare the distance between the discursive and material practices of state-making.

Figure 3.12 Road crew, Mocoa-Puerto Asís section, 1928.
Source: Archivo de la Diócesis de Sibundoy.

However, beyond the road's everyday drama what this story reveals is a structural element within the spatial history of the frontier. For it is no accident that, despite the expectations of wealth and civilization it aroused and the relentless pleas of missionaries and *colonos*, the road remained largely neglected, except when its existence seemed vital to preserve the physical integrity of the country. Fray de Montclar himself expressed this situation well in one of the many letters he sent to Bogotá questioning the government's inattention towards the road. In referring to the events that followed the episode of La Pedrera in 1911, he declared that:

> Once the danger was warded off, the armed force was removed from Puerto Asís, and the government lost interest in the conclusion of the ten leagues of road needed to reach a navigable point in the Putumayo River, and neither displayed great commitment in solidifying and building the bridges over the many rivers crossed by the road.[93]

A few years later, when the Apostolic Prefect had already left the Putumayo and resigned his post as Apostolic Prefect, Fray Florentino de

Figure 3.13 Fray de Montclar supervising the road works on the Mocoa-Puerto Asís section, 1927.
Source: Archivo de la Diócesis de Sibundoy.

Barcelona, the then director of road works, privately expressed a similar opinion in the following fragment of a letter sent to a friar colleague:

> My supreme ideal would be to be able to deliver the Umbría-Asís section decently, subject of two contracts with the Government, since from Umbría to Pasto we could say that [the road] had been in use for almost twenty years, and some benefit might have accrued to the territory, at least to allow the entrance of people, even though they are now not able to exit; in that way colonization is consolidated, keeping them in a jail or prison.[94]

At first glance, Fray de Barcelona's sarcastic statement seems like a conspiracy delusion of a friar exhausted by vainly struggling against a remote central government. Indeed, it is difficult to imagine that the road could have intentionally been conceived as a non-return way to the Putumayo. Besides, there is no reason to believe that this project was not intended to fulfil in practice the ends it symbolized in theory and, although in a very precarious and unstable way, parts of those ends were accomplished. The road became the main access route for the *colonos* from Nariño and other departments that gradually populated the Putumayo foothills and lowlands, assisting the Capuchin's goal of extending their evangelization activity across the region, and was instrumental in the government's imperative need of securing the country's territorial sovereignty. Yet, in another sense, the Capuchin priest's metaphorical association of the Putumayo with a prison, together with the numerous telegrams of *colonos* to the government expressing their feeling of confinement and neglect, cannot better describe the relationship of inclusive exclusion so pervasively present in the spatial history of the frontier.

This association brings us back to the notion of state hegemony discussed at the beginning of the chapter. As it was shown, the history of the Putumayo road was marked by countless conflicts, disputes and struggles, which together expose the gap between the rhetoric and practices of statecraft. And yet, this is a history that despite conflict and struggle speaks of continuity and perpetuation. Fray de Barcelona's image of the Putumayo as a space of confinement is enlightening precisely in this sense because it speaks of the immutability of the image itself. This is an image that reminds us of Reyes 30 years earlier when he saw this same region as a perfect place of exile for his political enemies; as Reyes' image reminds us of Fray de Montclar's early portrayal of the large territory entrusted to him as an incommensurable and isolated jungle disputed between the jaguar and the 'cannibal' Indians; as Fray de Montclar's portrayal reminds us of Codazzi's judgment of the *Territorio del Caquetá* as a place as 'backward' and 'savage' as the 'new world' Columbus encountered; and so on.

All these characters conceived plans and projects – roads, railroads, navigation schemes, 'white' immigration, evangelization– to 'civilize' and 'develop' the Amazon frontier and its 'inexhaustible' and 'unexploited' riches. At the same time, however, all these plans and projects depended on this large region remaining a frontier space or, in other words, retaining its condition of 'savagery' and 'backwardness' which defined its terms of inclusion into the state and sustained both the latter's civilizing fiction and the violence underlying the fiction.

Following Gramsci's notion of hegemony as a form of domination that goes beyond coercion, Lefebvre (1991) conceived space as an element instrumental within the production and reproduction of power relations in society. The ways in which Colombia's Amazon was discursively conceived and materially constituted as a frontier territory, as illustrated in this and the previous chapters, provides a clear example of how hegemony works through the production and maintenance of certain spaces and spatial relations. In reconstructing the history of the Pasto-Puerto Asís road, I sought to describe the characters, conflicts and practices through which this internal frontier was crafted. As shown, this is a history largely shaped by violence, a violence that expresses itself not just in physical ways but in a wide array of other, more subtle forms: in the production of visual and cartographic silences, in the establishment and constant reinforcement of spatial and racial tropes, and in the institution and institutionalization of certain language.

It was through the normalization of these silences, tropes and language, through its embeddedness in time and space, that the state–frontier relationship came to appear as the natural order of things. In the remaining chapters, I will look at how this relationship continues to manifest and reproduce itself within the contemporary space of the frontier, and also at how it is made sense of and challenged in everyday life.

Notes

1   There are several excerpts of Gramsci's Prison Notebooks where the author emphasizes this dialectical or relational character of hegemony (e.g. Gramsci 1971, p. 52, 56, 182). It is important to mention that in Gramsci's analysis relational does not necessarily mean equivalence or symmetry, and hegemony is primarily seen as a process by which certain class ideologies and values become dominant or end up prevailing in society. For a similar reading of hegemony see Jessop (2009, pp. 101–117) and Roseberry (2004).

2   Some ethnographical accounts of the state have examined in different ways the agency of nature within the context of state-making. See, for instance Harvey (2005, 2001) and Scott (2009).

3 'Resolución No.264'. Gobernación Departamento, Pasto, 17th November 1909, APCC, Missions Caquetá-Putumayo, Box 4, n.f.

4 'Resolución No.264'. Gobernación Departamento, Pasto, 17th November 1909', APCC, Missions Caquetá-Putumayo, Box 4, n.f.

5 See, for example: AGN, MOP, Vol.1408, fols.410–413v; Vol.1409, fol.375v; Vol.1413, fols.30–33.

6 'Informe de los trabajos del camino del Oriente', 30th October 1910, AGN, MOP, Vol.1407, fol.323.

7 'Informe que rinde el ingeniero comisionado por el Ministerio de Obras Públicas para verificar una inspección al camino al Putumayo', October 1913, AGN, MOP, Vol.1409, fol.311.

8 Manuel Silva to Estanislao de Las Corts, 20th September 1912, ADS, folder 11-010-01, Camino Pasto-Puerto Asís asuntos varios, n.f.

9 Rogerio Becerra to Sobrestantes y Caporales del Camino Nacional, 4th November 1912, ADS, folder 09-01, Camino al Oriente, n.f.

10 Vicente Narváez to Minister of Public Works, 15th May 1914, AGN, MOP, Vol.1409, fol.576v.

11 The most widely known work on this subject is the one published in 1968 by the Colombian anthropologist Victor Daniel Bonilla (see Bonilla 1972). See also Casas (1999) and Gómez (2011, 2005).

12 'Proyecto de Colonización del Caquetá y Putumayo', 15th September 1911, AGN, MOP, Vol.1407, fols.417–421.

13 For a detailed account of the foundation and early history of Puerto Asís see Revelo (2006), Vilanova (1947, vol.1, pp.276-287) and 'Puerto Asís y el pueblo Colombiano', 21st November 1914, ADS, folder 9-II 10-03, n.f.

14 The *cuadra* or *fanegada* is a unit of land measure equalling 6,400 square metres.

15 Estanislao de Las Corts to Apostolic Prefect, 7th May 1912, quoted in Vilanova, *Capuchinos Catalanes*, Vol.1, p. 282 (emphasis added).

16 Fidel de Montclar to Andrés de Cardona, 19th April 1912, ADS, folder 11-010-01, Camino Pasto-Puerto Asís asuntos varios, n.f.

17 *Las Misiones en Colombia*, p. 137.

18 Fray de Las Corts to Fray de Montclar, 16th September 1912, ADS, folder 09-01, Camino al Oriente, n.f.

19 Velasco to Minister of War, AGN, MRE, Diplomática y Consular, Box 742, folder 323, fols.36-36v.

20 In legal terms, *baldíos* are lands owned by the state with the purpose of awarding them to individuals who comply with certain requirements established by the law.

21 'Ley 52 de 1913 (5 noviembre). Por la cual se provee la colonización del Caquetá y Putumayo', APCC, Missions Caquetá-Putumayo, Box 1, n.f.

22 'Garantías y condiciones para los que desean establecerse en la Colonia de Puerto Asís', 8th July 1913, ADS, Colonia Puerto Asís, n.f.

23 'Prospecto para la colonización del Putumayo', 20th April 1914, APCC, Missions Caquetá-Putumayo Box 3, n.f.

24 Although the *Junta* had been assigned executive functions, the regular newsletters issued by this board suggest that its role ended up being rather

limited and largely restricted to spreading and publicizing the colonization projects region wide. See, for example: 'Boletín de La Junta de Inmigración Nos.1-2', 20th July 1914, Pasto, Imprenta del Departamento, in APCC, Missions Caquetá-Putumayo, Box 1, n.f.

25   n.t., 20th January 1927, APCC, Box 3, n.f.

26   See, for instance: Fray de Montclar to Minister of Public Works, AGN, MOP, Vol. 1414, fols. 415–421; and Montclar (under the pseudonym Delfín Iza, Iza 1924, pp.2–4, 13–15).

27   *Colonos* Puerto Asís to Minister of Public Works, 21st January 1926, AGN, MOP, Vol.1414, fol.351.

28   This extension corresponds to the flat area of the Sibundoy Valley, whose total area – including the surrounding mountain zone – has been estimated at 52,000 hectares. However, during most of the twentieth century a considerable part of the flat zone – or the valley properly speaking – was marshland.

29   Rufino Gutiérrez to Ministry of Public Works, 24th June 1912, AGN, MOP, Vol.1407, fols.579–582.

30   Rufino Gutiérrez to Ministry of Public Works, 21st April 1912, AGN, MOP, Vol.1407, fols.561–562v.

31   For a general history of this process see the excellent work by Legrand (1986). For the specific context of the Amazon and Putumayo regions see, amongst others: Domínguez (1984) and Fajardo (1996).

32   'Ley 51 de 1911. Por la cual se ceden unos terrenos baldíos al Departamento de Nariño y se manda a fundar una población', 18th November 1911, APCC, Box 1, Missions Caquetá-Putumayo, Box 1, n.f.

33   'Ley 106 de 1913. Que adiciona y reforma la 51 de 1911 y ratifica una cesión de terrenos baldíos', 29th November 1913; 'Ley 69 de 1914. Por la cual se reforman las leyes 51 de 1911 y la 106 de 1913', 12th November 1914, APCC, Box 1, Missions Caquetá-Putumayo, Box 1, n.f.

34   A map of the Tamabioy's lands and a transcription of the referred will can be found in Bonilla (2006).

35   The Indians referred here specifically to Law 89 of 1890, which contemplated a series of measures to safeguard the indigenous property rights over *resguardos*.

36   Indians from Santiago to President, 31st July 1914, APCC, Missions Caquetá-Putumayo, Box 4, n.f. (underline in original); see also Francisco Tisoy e hijos to President, 13th October 1913, AGN, MG (4th), Vol.74, fol.52.

37   Sucre changed its name to Colón in 1936.

38   'Crónica Misional del Putumayo (1893–1968)', ADS, Miscellany, n.f.

39   Fidel de Montclar to Ministry of Public Works, 15th August 1920, AGN, MOP, Vol.1412, fols.458–468.

40   N.t., 25th March 1922, APCC, Missions Caquetá-Putumayo, Box 21, n.f.

41   N.t., 25th March 1922, APCC, Missions Caquetá-Putumayo, Box 21, n.f.

42   Fidel de Montclar to Ministry of Public Works, 1st December 1911, AGN, MOP, Vol.1407, fols.467–470.

43   Fidel de Montclar to Ministry of Public Works, 20th February 1912, AGN, MOP, Vol.1407, fol.526v.

44 Estimation based on the average time responses of letters, reports and tele-
    grams from 1906 to 1932.
45 Fray de Montclar to Fray de Cardona, 1st March 1912, ADS, folder 09-01,
    Camino al Oriente, n.f.
46 AGN, MOP, Vol.1407, fol.524.
47 AGN, MOP, Vol.1407, fol.525v.
48 There are several cases illustrating this. See, for example: Fray de Montclar
    to Fray de Cardona, 1st January 1913, ADS, folder 11-010-01, Camino
    Pasto-Puerto Asís, asuntos varios., n.f.; Peones del Camino to Minister of
    Public Works, 28th April 1913, AGN, MOP, Vol.1408, fols.463–464, 549;
    Fray Florentí de Barcelona to Fray Canet del Mar, 29th October 1930,
    ADS, folder 11-10-01, Camino Pasto-Puerto Asís, n.f.
49 Fray de Las Corts to Minister of Public Works, 7th July 1910, AGN, MOP,
    Vol.1407, fols.297–299v.
50 Fray de Las Corts to Minister of Public Works, 30th October 1910, AGN,
    MOP, Vol.1407, fol.324.
51 'Decreto No.426, noviembre 9 de 1991. Que contiene algunas disposicio-
    nes reglamentarias y de Policía sobre caminos', Pasto, Imprenta del Depar-
    tamento, AGN, MOP, Vol.1408, fols.353–356v.
52 AGN, MOP, Vol.1407, fols.430v, 447.
53 Guerrero to Ministry of Public Works, July 1912, AGN, MOP, Vol.1407,
    fols. 589 -590v, 604–605.
54 Fray Canet del Mar to Minister of Public Works, 4th November 1922,
    AGN, MOP, Vol.1413, fols.30-33.
55 'Carga transportada y vehículos que han recorrido la vía de Pasto al Putu-
    mayo durante el periodo de 1 de marzo de 1922 a 30 de abril de 1923',
    April 1923, AGN, MOP, Vol.1413, fol.94v.
56 Mora to Minister of Public Works, 13th November 1923, AGN, MOP,
    Vol.1413, fols.187–188.
57 Indígenas de La Laguna to Presidente de la República, 27th March 1924,
    AGN, MOP, Vol.1413, fol.250.
58 'Informe que rinde el ingeniero comisionado por el MOP para verificar una
    inspección al camino del Putumayo', AGN, MOP, Vol.1409, fols.327–328.
59 Fray de Montclar to Fray de Cardona, 21th June 1912, ADS, folder
    11-010-01, Camino Pasto-Puerto Asís asuntos varios, n.f.
60 'Informe del Comisario de Mocoa al Ministro de Gobierno en el que in-
    forma de sobre el estado del camino', 1st July 1912, AGN, MG (4th),
    Vol.68, fols.303–306.
61 Roberto Salazar to Minister of Public Works, 24th August 1912, AGN,
    MOP, Vol.1408, fols.48–53.
62 'Resolución por la cual se mandan a organizar los trabajos de conservación y
    mejora de la parte construida del camino que conduce de Pasto hacia el río
    Putumayo, por Mocoa', 30th August 1912, AGN, MOP, Vol.1408, fols.9–11.
63 Fray de Montclar to Minister of Public Works, 3rd September1912, AGN,
    MOP, Vol.1408, fols.19–20.
64 Fray de Las Corts to Fray de Montclar, 1st September 1912, ADS, folder
    09-01, Camino al Oriente, n.f.

65 Fray de Las Corts to Fray de Montclar, 17th September 1912, ADS, folder 09-01, Camino al Oriente, n.f.

66 Thomas to Minister of Public Works, 20th October 1912, AGN, MOP, Vol.1408, fols.173–178v.

67 Fray de Montclar to Presidente de la República, 7th November 1912, AGN, MOP, Vol.1408, fol.116.

68 'Informe de la Comisión nombrada por el Gobierno Departamental de Nariño para inspeccionar la vía' 1912, Imprenta del Departamento, Pasto, AGN, MOP, Vol.1408, fols.299–313v (emphasis added).

69 Escandón to Minister of Public Works, 5th December 1912, ANG, MOP, Vol.1408, fol.160; See also fols.236–258.

70 Fray de Las Corts to Fray de Cardona, 2nd January 1913, ADS, folder 09-01, Camino al Oriente, n.f.

71 'La Suspensión del camino y sus motivos', Marzo 1913, Imprenta de la Diócesis, Pasto, APCC, Missions Caquctá-Putumayo, Box 1, n.f.

72 'Resolución que reorganiza los trabajos del camino al Putumayo', 14th March 1914, AGN, MOP, Vol.1409, fols.505–506.

73 On the conflict between Thomas and the Capuchin Mission see: 'Defensa de los R.P. Capuchinos y contestación a un informe calumnioso de don Julio Thomas', 30th April 1913, AGN, MOP, Vol.1408, fols.579–589; AGN, MOP, Vol.1408, fols.549, 562–564v, 599; 'Injusticias en el Putumayo', Nuevo Tiempo, 27th June 1913; APCC, Missions Caquetá-Putumayo, Press, n.f.; 'Sección Misiones', Deber No.86, 15th April 1914, APCC, Missions Caquetá-Putumayo, Press, n.f.

74 'Contrato celebrado en licitación pública con el Sr. Vicente Micolta C., para la mejora, terminación y conservación del camino de Pasto a Puerto Asís', 13th September. 1915, AGN, MOP, Vol.1411, fols.391–393v.

75 Fray de Montclar to Minister of Public Works, 13th March 1917, AGN, MOP, Vol.1411, fols.487–490.

76 'Visitaduría Fiscal', 12th August 1917, AGN, MOP, Vol.1411, fol.578.

77 'Contrato por el cual se resuelve lo relativo al camino del Putumayo', 6th December 1917, AGN, MOP, Vol.1911, fols.660–662v.

78 Márquez to Minister of Public Works, AGN, MOP, Vol.1412, fols.43–46.

79 Fray de Montclar to Minister of Public Works, 6th May 1918, AGN, MOP, Vol.1412, fols.63–68.

80 AGN, MOP, Vol.1412, fols.60–62, 129, 163–163, 173–177.

81 León to Minister of Public Works, 16th June 1919, AGN, MOP, Vol.1412, fol.146v.

82 'Resolución No.11, Por la cual se declara administrativamente caducado un contrato', 29th May 1924, AGN, MOP, Vol.1413, fol.294.

83 'Algo por la verdad', El Nuevo Tiempo No.7719, 4th July 1924; 'El Camino al Putumayo', El Nuevo Tiempo No.7708, 23rd June 1924, APCC, Missions Caquetá-Putumayo, Press, n.f.

84 'Decreto ejecutivo No.213. Por el cual se reorganizan los trabajos de Pasto a Puerto Asís', 10th February 1925, AGN, MOP, Vol.1414, fol.32.

85 See: ANG, MOP, Vol.1414, fols.115–119, 351–355, 415–421, Vol.1415, fols.279–283, Vol.1440, fols.120–121; Vecinos de Puerto Umbría to President of Senate, 9th August 1926, APCC, Missions Caquetá-Putumayo, Box 3.

86 Fray Canet del Mar to Ministry of Public Works, 28th December 1928, AGN, MOP, Vol.1415, fols.424–426.

87 Vecinos de Mocoa to Ministers of Works, Industry, Associated Press, 11th March 1929, ANG, MOP, Vol.1415, fols.437–438.

88 See, for instance: Vecinos de Mocoa to President, 26th February 1929, AGN, MOP, Vol.1440, fols.131–134; Vecinos de Sucre to Ministers of Government, Industry, Treasure, Public Works, 12th July 1919, AGN, MOP, Vol.1415, fols.469–471; Fray Canet del Mar to Minister of Public Works, 10th May 1929, AGN, MOP, Vol.1415, fols.289–291.

89 'Decreto 1549. Por el cual se dispone la ejecución de algunos trabajos', 26th September 1932, AGN, MOP, Vol.3273, fols.1–3; 'Decreto 1581. Por el cual se fija el plan de trabajo que debe adoptarse en algunas vías', 28th September 1932, AGN, MOP, Vol.0318, fols.8–11; AGN, MOP, Vol.3278, fol.36.

90 Rafael Agudelo to Minister of Public Works, 11th December 1932, AGN, MOP, Vol.3273, fols.199–214, fol.210.

91 'Documentos referentes a la muerte del Rvmo. Padre Fidel de Montclar. Acaecida en el Convencto de Arenys de Mar (España) el 21 de marzo de 1934', ADS, Fidel de Montclar, n.f.

92 'El célebre misionero de la Amazonia Colombiana', *El Derecho* No.754, 24th March, 1934, ADS, Fidel de Montclar, n.f.

93 Fray de Montclar to Minister of Public Works, 16th April 1926, AGN, MOP, Vol.1414, fols.415–421.

94 Fray Florentino de Barcelona to Fray Canet del Mar, 4th November 1930, ADS, folder 11-10-01, Camino Pasto-Puerto Asís, n.f.

Part II

4

The *trampoline of death*

Discipline is essentially centripetal … discipline functions to the extent that it isolates a space, that it determines a segment. Discipline concentrates, focuses, and encloses. This first action of discipline is in fact to circumscribe a space in which its power and the mechanisms of its power will function fully and without limit

(Foucault 2007, pp.44–45).

Whether we assume the state as the primal *source* of power or the *effect* of governmental technologies, Michel Foucault's remark on discipline closely resembles the frontier-state relationship described in the previous chapters. In trying to elucidate the nature of this relationship, I have drawn on the notion of the state of exception, a notion that, as observed by Agamben, allows us to recognize how the validity and legitimacy of the juridico-political order of the state ultimately depends on the delimitation of a space 'in which it is possible to trace borders between inside and outside and in which determinate rules can be assigned to determinate territories' (Agamben 1998, p.21).

There are three interrelated arguments that have so far been advanced and stem from this conceptual framework. Firstly, the frontier does not constitute a territory excluded from the order of the state but a space (and more broadly a *condition*) constituted through a topological relation of inclusive exclusion, that is, as previously mentioned, 'the extreme form of relation by which something is included solely through its exclusion' (Agamben 1998, p.18). Secondly, this relation, whose origins in the context of the Amazon we can trace back to the colonial discursive

Frontier Road: Power, History, and the Everyday State in the Colombian Amazon, First Edition. Simón Uribe.

and material practices of domination and control, is immanent within the foundational myth of the modern (postcolonial) state. And thirdly, as long as this myth is founded and sustained on a series of oppositions or antinomies – civilization vs. savagery, nature vs. culture, centre vs. periphery – this relation is by definition indissoluble. Put in a different way, although state and frontier often appear as two irreconcilable orders, the latter represents not an obstacle but a condition of possibility in which the former's legitimacy and power is rooted and maintained.

In describing how this relation originated and became hegemonic, I stressed the importance of seeing how the state functions simultaneously as a discursive and material force, and how these two dimensions of power have shaped the spatial history of the frontier. The pervasive image of the frontier as an 'exceptional' space in its different connotations or expressions ('no man's land', 'prison', 'savage territory', 'terra incognita', and so forth) referred to in the previous chapters is illustrative in this respect, because it reveals how state hegemony ultimately depends not on the failure or success of the many projects and plans by which the state attempts to dominate and control a given space or population but on how *through* them this image persists and is constantly reinforced. However, as I have attempted to show, frontiers are not just a rhetorical projection of the state or a passive locus of power but concrete (material, social and cultural) spaces where the state is encountered or manifested in and through a wide variety of characters, places and relations. This requires conceptualizing the frontier–state relationship in terms of both the transformative (creative-destructive) effects it produces, as well as the different responses it elicits.

This chapter, and the two that follow, continue to examine this relationship. I now turn from the early history of the road to an ethnographic exploration of different contemporary instances, characters and events that cast light on frontier people's material and affective relations with this exceptional infrastructure. In this chapter, I am especially concerned with how people at the frontier make sense of some of the *perceived* and *lived* realities that the road symbolizes: abandonment, neglect, isolation, death, confinement, anger, fear. Making sense of these realities always involves an act or a process that is simultaneously historical and spatial. For most inhabitants of the Putumayo, to talk about the road is to talk about a space that is saturated with historical events and memories through which they imagine and locate themselves in relation to the state: past and present struggles, unfulfilled promises and expectations, enduring feelings of exclusion and desires of inclusion. In the same way, these events and memories are deeply imprinted and only fully make sense in the everyday intricate geography of the road: its hairpins and precipitous curves, broken surfaces, landslides, police and army checkpoints, scattered shrines and crosses.

But making sense of the road is also, and fundamentally, a political practice. By locating, tracing and connecting those memories and events in *time* and *space*, history and geography are inexorably invested with a moral content through which people situate and imagine themselves and others as moral subjects.[1] Through this moral geography and history, the rhetoric of 'progress', 'civilization' and 'modernity' that the road so powerfully embodied but never fulfilled is appropriated and turned into an array of economic, social and political universal claims not only about people's past and present but their probable futures. The main paradox of such claims is perhaps that they often contest certain forms of exclusion or inclusion whilst at the same time they reproduce others. The characters introduced in this chapter draw attention to this paradox, specifically in the way in which they produce or legitimize narratives that privilege some histories, characters and events whilst silencing others. In other words, these narratives speak of how hegemony is both contested and reproduced in the space of the frontier.

The chapter is presented in three parts. The first takes up the history of the Capuchin Mission's road in its last years and relates how it was eventually transformed to a road suitable for motor vehicles, emphasizing the central role it played in the region's history. This is a story of promise and despair, emerging both from the redemptive power that roads were imagined to have in the life of the frontier and the many dramas they involved in practice. The road fully exemplifies this story and has become with time one of the most pervasive symbols through which the inclusive-exclusive relationship between state and frontier is exposed. The second part is built around a series of individual narratives about the road and seeks to cast light on how particular people make sense of this infrastructure at the different levels mentioned: spatial, historical and political. Although each narrative might emphasize any of these levels, they all relate to the three in different ways. In the final part, I return to questioning how these narratives represent both forms of contestation and reproduction of this relationship, and stress the significance of this twofold dimension in the spatial history of the frontier.

A frontier highway

The construction of the land route connecting the Andean city of Pasto with the port town of Puerto Asís took a quarter of a century to complete. Rafael Reyes, who had first envisioned the road in the 1870s, died in 1921, 15 years after the Capuchin missionaries had begun its construction and ten years before it finally reached its destination. Fray Fidel de Montclar, who fervently embraced the project and for more than two decades struggled continuously against its countless obstacles, left the Putumayo without seeing it finished. Moreover, and an ironic epitaph for this calamitous project, by the time it was finally completed, it was unable to meet transport needs, a situation that would soon become evident during the armed conflict with Peru. Still, the story of how the three-metre wide 215-kilometre bridle path that brought so many eulogies, travails and nightmares to the Apostolic Prefect and his fellow missionaries became a 'modern' motor road is as long and dramatic as the story of the Capuchin road.

As early as 1925, when a significant chunk of the road to Puerto Asís was still unfinished, a national law decreed the upgrade of its first 25 kilometres – from Pasto to La Cocha lake – to a *carreteable* (a motor road). Two different teams of engineers were contracted between 1926 and 1927 to carry out the layout design, and construction works finally began in December 1927.[2] Between that date and July 1928, according to a report submitted by Jeremías Bucheli, chief engineer of works, the layout was completed and 5.2 kilometres of road had been built. Although in some parts the new road followed a different route from the old one, the works largely consisted of the widening (to eight metres, including ditches) and improvement of the latter's surface.[3] Apparently, things were initially going well and the main problem thus far, as reported by Bucheli, had been that some land owners along the route were demanding too high a compensation rate for the road to be built on their property. However, during 1928 the budget was reduced significantly and then, in February 1929, the central government suddenly issued a decree ordering the suspension of the works claiming fiscal constraints.[4]

Almost four years later, in December 1932, the road was abandoned and had not advanced an inch beyond the five kilometres built by Bucheli.[5] A year earlier – through Law 88 of 1931 – it had been incorporated into the national road network and its extension to Puerto Asís ordered, but this mandate had remained on paper. The conflict with Peru broke out in September 1932 exposing, once again, the vulnerable condition of the country's frontier regions. The Pasto-Puerto Asís road was the only land access route to the Putumayo River, yet its precarious state constituted an obstacle even for the movement of infantry troops. Thus, and in the rush of war, a series of last-minute decrees were passed to overcome this critical situation. One of them commanded the immediate 'construction, maintenance and improvement of those roads that the government deemed necessary to adequately meet the protection of the national frontier in the south of the country'.[6] Another, issued a few days later, catalogued the Pasto-Puerto Asís road as the highest priority and ordered its immediate repair.[7]

Three months later, on 11th December 1932, Rafael Agudelo, the engineer in charge of the first road section (Pasto to Puerto Umbría) reported that the entire 160 kilometres of road had been totally repaired. However, in the same report, he insisted to the Minister that keeping the existing road would be useless and asked him to consider its total upgrade. A new gravel road, noted the enthusiastic engineer, would guarantee not only the country's territorial sovereignty but 'a true extension of the national territory and an immeasurable increase of public wealth'.[8] The answer came quickly. On 21st December, Agudelo received a letter informing him that 'for reasons of economy', his position would have to be terminated and the section under his charge reallocated between the engineer responsible for the last section of the road (Puerto Umbría to Puerto Asís) and the Pasto road zone engineer. With regards to his eager appeals for the *carreteable* – in another report he had gone as far as to assure the Minister that with 2,000 workers the whole road could be improved in four months[9] – they were simply ignored.

The road had to wait a further three years before the government finally hired a new engineer to resume the upgrading works, frozen since 1929. The new contract contemplated the upgrade of the road between the few kilometres built by Bucheli and the site of Urcusique (16 kilometres south from Mocoa) and divided the road in two sections.[10] The first (Pasto-San Francisco) was completed within the agreed terms and was put into service by the end of 1936. The latter was initiated that same year and was expected to be concluded in nine months. However, it wasn't until eight years later, and after countless conflicts, bureaucratic hurdles and pleas of *colonos* and missionaries, that the road arrived at Mocoa. Yet it took another 13 years to reach its end point on the Putumayo River. Its arrival at Puerto Asís in November 1957 is remembered as being a major landmark in the town's history. A historical brochure published in 1961 and entitled 'Puerto Asís ayer y hoy' [Puerto Asís yesterday and today] recalled the crowded Mass and parade that took place on the occasion and described the joy of the people as 'indescribable'. The event, it added, 'marked a new age of development and intensified considerably the colonization along the road as well as its surroundings' (Plácido 1961).

The construction years of the *carreteable* from Pasto to Puerto Asís coincide roughly with a shift in the government's transport policy from railroad to road construction. As a result, between 1930 and 1950 the country's road network more than tripled, from less than 6,000 kilometres to around 21,000 kilometres. Nevertheless, and despite this significant growth, by the early 1950s Colombia had less than two kilometres of roads per 1,000 inhabitants, the lowest per capita road length in Latin America (Pachón and Ramírez 2006, pp.54–55). The quality of roads was also among the poorest of the region, with less than 5% of the total national road network paved: the equivalent of less than one kilometre per 1,000 inhabitants. The country's difficult topography, the government's financial limitations, the high maintenance costs and the prevalence of regional interests over national ones are some of the reasons adduced for this lag. More striking, however, was the high concentration of the road network in the country's central departments and its virtual absence in frontier regions: of the total 18,500 kilometres of roads in 1945, only 613 kilometres (none of them paved) were located in the *Intendencias* (Intendancies) and *Comisarías especiales* (special commissionerships), the peripheral territorial units that together comprised about three-quarters of the country's area. Beyond the obvious economic and demographic rationale underpinning this disparity, this pronounced 'road inequality' gap – a situation that remains largely unchanged today – cannot but remind us of the continuing frontier condition of these territories.

The scarcity and precariousness of frontier roads did not deter optimism in terms of their beneficial effects for the country. Heavily influenced by early frontier scholars such as Frederick Turner, Isaiah Bowman and Herbert Bolton, a generation of geographers saw these infrastructures traversing the steep Andean mountains and plunging into the vastness of the llanos and jungles as a powerful redemptive technology of the poor man's quest for land and domination of the wilderness. Raymond Crist, a US geographer who visited Colombia in the late 1940s under the auspices of the Smithsonian Institution, deemed road development in frontier regions as a revolutionary change in the country's history. 'Such roads', commented Crist in a book about frontier settlement experiences in South America, 'with their jeeps, cars, and trucks, represent a kind of safety hatch for those who will and can leave the Middle Ages, either for the twentieth-century life in towns or cities or for the unpopulated stretches of the east' (Crist and Nissly 1973, p.iv). Ernesto Guhl, Crist's research assistant during his travels in the country and a prominent figure in Colombian geography, devoted several pages of his classic work *Colombia: bosquejo de su geografía tropical* [Colombia: an outline of its tropical geography] to praise the economic and cultural changes brought about by the boom in road construction. When referring to the advance of colonization in the Amazon and Llanos frontiers, he listed the road along with the river and airfield as the 'three factors' preventing such settlements becoming 'a colony of exile' (Guhl 1976, vol.2, p.130).

But the most fervent exaltation of frontier roads can be found in the work of another of Crist's disciples: Edmund Hegen's *Highways into the Upper Amazon Basin*, a case study from the early 1960s of six 'penetration roads' to the Amazon regions of Colombia, Ecuador and Peru. Pregnant with Ratzelian overtones, Hegen's book portrays those roads as the pioneer settler's ultimate instrument for his everyday struggle of turning the inhospitable landscape of the jungle into a cultured 'living space' or *Lebensraum*. A typical passage reads:

> In the Selva, man has only two choices. The first is to accept the rules of the forests, the rivers, and rains. This means small space, narrow resource base and isolation … The other choice is to confront the forest with man's supreme tool – the predetermined road, the lifeline of the cultural landscape. From this break in the front of the forest, man will wrest the space of the cultural landscape (Hegen 1966, p.111).

This over-enthusiastic view of the frontier road, which Hegen tellingly illustrated through the Pasto-Puerto Asís 'highway', could hardly be sustained today. Not only does this image scarcely match what most of those highways really were – *trochas* or narrow one-way dirt roads impassable during rainy seasons but the 'frontier experience' in general, as discussed, was anything but idyllic and barely promoted egalitarian and democratic values, as Turner famously argued in the context of North America. Still, the significance of roads within the history of frontiers can hardly be over-emphasized. Although more precarious 'escape valves' than 'highways' or palimpsests smoothing the advance of 'civilization' over 'nature', such roads actively supported and often shaped the temporal and spatial patterns of colonization and settlement, as well as the different dynamics linked to them. This is in many ways the case of the Amazon's twentieth-century colonization processes, for the most part distinguished by its spontaneous nature or the lack of planned schemes and government support. This situation was clearly stated as early as 1945 by the Colombian ethnologist Milciades Chaves, who concluded that colonization in the region of Putumayo had primarily been 'carried out by hunger, without any interest on the part of the State or official institutions'. The *colonos* – mostly impoverished and landless peasants from the department of Nariño and to a lesser extent from Cauca and Huila – he added, 'have so far been abandoned to their fate, though they constitute a productive force' (Chaves 1945, p.580). As for the indigenous peoples, their situation was, according to Chaves, the most distressing one, constantly displaced and violently dispossessed of their lands.

The bleak picture painted by Chaves in the early 1940s, as many later studies have repeatedly stressed, has remained largely unaltered over time (e.g. Casas 1999; Domínguez 1984; Gómez 2011; Ortiz 1984). In the late 1940s and 1950s, peasants fleeing from 'La Violencia', a bloody partisan conflict that spread over many of the country's rural areas, migrated to the Putumayo following the Pasto-Puerto Asís road. However, the bulk of the new settlers continued to be landless peasants coming mostly from the neighbouring provinces or floating populations attracted by the booms or extractive cycles that together tell an important part of the region's contemporary history: timber since the 1950s, fur in the 1960s, oil during the 1960s and early 1970s, *coca* since the 1980s and, more recently, since the late 2000s, mining and again oil. Just from a demographic standpoint, the impact of such migrations is noteworthy. Between the 1950s and the 1990s, the four decades during which most of the mentioned cycles occurred, the population of the Putumayo increased more than sevenfold, from 28,105 according to the 1951 census to 204,309 in 1993 (Ariza, Ramírez and Vega 1998, p.184).[11]

The role the road played in this process was vital. The colonization and settlement patterns in the Putumayo have been tellingly described a 'linear' or 'linear disperse' (Domínguez 2005, p.260) since they largely followed the course of the 220-kilometre dirt road and the banks of the Putumayo River and some of its main tributaries (Figure 4.1). The occupation of lands adjacent to the *carreteable*, a phenomenon that dates

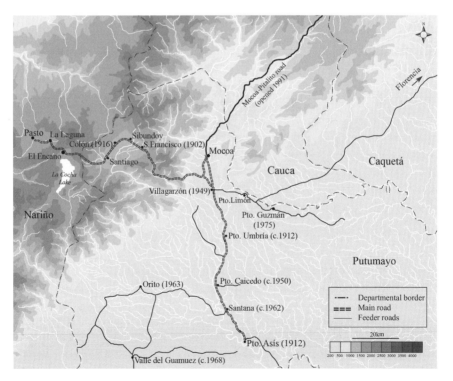

Figure 4.1 Towns along the *carreteable* Pasto-Puerto Asís and feeder roads. Source: elaborated by author.

back to the time of the Capuchin road, has been explained in both economic and geographical terms. Economically, because only agricultural activities in those lands with easy access to transport routes could be considered profitable (Domínguez 2005, p.260); geographically, because the strip of flat land between the road and the *cordillera* to the west is relatively narrow, and the access to the road and the Putumayo river – which runs parallel from Puerto Umbría to Puerto Asís – from the east was difficult due to the lack of tributary rivers flowing in that direction (Brücher 1968, p.151). Yet, as noted by Rolf Wesche, who extensively surveyed the region in the early 1960s, the significance of the road in the region's colonization process was not confined to its geographic layout or economic benefits. The road, together with the main rivers, he observed, '[constitute] a place of excursion, the route through which news travel, and the primary means of social contact'. Whilst noting that 90% of the Putumayo's inhabitants lived at a maximum distance of one kilometre from those arteries, the German geographer stated that a distance of just 100 metres from them 'could represent the difference between taking active part in daily life or total isolation' (Wesche 1974, pp.1–2).

Reyes' ghost

If one tries to imagine how the 'frontier experience' might look from the perspective of the 'pioneer', Wesche's emphasis on the significance of the road can be easily appreciated. Roads are widely assumed as synonymous with mobility and access to markets, public services and jobs. Yet for the *colono*, who had left behind his native land and adventured through those 'penetration roads' in the hope of a better future or livelihood, these infrastructures, no matter how precarious they were, meant something more. Faced with the harsh everyday realities of frontier life – lack of government support, remoteness, confinement – the road for the *colono* meant a vital link, both physical and imaginary, with the world beyond the isolated space of his farmland. The significance of this link, moreover, is further accentuated if we consider that in the Putumayo the *trocha* from Pasto to Puerto Asís constituted for several decades the only land route connecting this part of the Colombian Amazon to the rest of the country.

Even today, half a century after Wesche visited the Putumayo, and despite the improvements in transport infrastructure, his remarks on the road remain in many ways valid. And yet there is a story about this specific road that makes it singular even among other frontier roads. This story began around 1932 with Rafael Agudelo, the same engineer contracted by the government to repair the road at the time of the conflict with Peru. As mentioned earlier, Agudelo insistently asked the Minister of Public Works about the urgent need to upgrade the entire infrastructure, a work that, according to him, consisted mostly of widening the actual road. However, he strongly argued for the need to totally replace the San Francisco-Mocoa section, of which he pointed out a number of major limitations and technical faults: the road gradient was too steep, the large number of rivers and creeks along the way required the construction of several bridges and culverts, and a long stretch of the road went through a section of solid rock hence making its widening an onerous task. To make his point even clearer, he employed the terms 'eyesore', 'irrational', 'anti-economic' and 'truly indigenous trail' when referring to the Capuchins' most praised accomplishment.[12]

Agudelo's proposal – a plan consisting of two new roads starting from the Sibundoy Valley with one ending in Mocoa and the other in Puerto Umbría – was never considered. Nevertheless, his sustained attack on the San Francisco-Mocoa section seems to have predicted its end. In 1936, when the upgrade works were finally contracted, the decision was to avoid the current bridle path and to build a new road following a different route. The construction of this road proved as turbulent and chaotic as that of the Capuchin road. The archive's documents shed light on the usual messiness of bureaucracy and everyday troubles.[13] Putumayenses, meanwhile, bring to memory strong echoes of thunderous explosions and lines of workers hanging from ropes carving the road's path through the precipitous *cordillera*. But the real calamity of the new road seems to have been not a 'tragedy repeated as farce' but a farce turned into tragedy. The verdict in 1942 by the engineer in charge of the works – the first one had long since resigned – does not require much clarification: 'The conditions of this section, especially regarding to road width, curvature and drainage works', he emphasized in a detailed report, 'constitute a threat to the lives of those travelling there … Commercially and even for national defence purposes, a case in which a solid and safe road is required, this sector will be useless'.[14] Neither does the answer from Bogotá a month later: 'the works, above all', reads a fragment of the short letter from the National Office of Roads and Railroads, 'must be aimed at connecting the sections currently isolated, allowing traffic while a definitive solution to this problem is sought'.[15]

Decades of tragic accidents, complaints, pleas, unfulfilled promises, strikes and funereal names with which this infamous road section has been baptized (of which the *trampoline of death* remains the most popular one), bear witness to the nefarious effects of ignoring the engineer's warning. Still, the biggest irony of this episode is perhaps that after several years of numerous technical studies it was eventually concluded that the best possible 'definitive solution' would be to build a new road that roughly followed the same route chosen by the Capuchin missionaries around 1910.

Why the route of the *trampoline of death* was chosen in the first place is however less clear. As early as 1939, an enraged citizen from Mocoa wrote directly to the President protesting not only about the road's absurd layout but about the fact that it avoided the capital of the province. In response, he got a cold note from the President's Secretary reprimanding him for the 'inappropriate tone' of his letter and pointing out that the current project was the result of 'long and meditated analyses'. The road, he stated, had not followed the Capuchin route 'for the simple reason that [this route] required the construction of 20 long bridges over the numerous rivers which crisscross that place'.[16] However, as if the damage could be easily repaired, he noted that a branch road connecting Mocoa with the main road was contemplated.

Although this branch road was eventually built, even today, more than 60 years later, old Mocoanos still regard this event as a deep affront from which the town could never totally recover. For them, moreover, the 'true' reason why the road did not pass through Mocoa lies elsewhere and dates back to the time of Rafael Reyes. Supposedly, or so the story goes, a relative of one of the engineers hired to build this section was among the political prisoners exiled to Mocoa by Reyes (Chapter 1), and for this very reason he had decided to take revenge on this town by choosing a route avoiding the province's capital.

I was told this story several times, although the name of the engineer and his relative often varied from one version to another. Neither was this the only thesis I heard about the origins of the *trampoline of death*: another I came across a couple of times argued that the road had been strategically designed to repel the eventual advance of the Peruvian army beyond Mocoa, thus evidencing the little interest the government had in protecting the Amazon. Whether these stories are factual or not is not the point here. What is interesting is how through them people make sense of an event that surfaces often when making claims about the region's past, present and future. It is to this subject that I now turn in order to illustrate how frontier peoples imagine and locate themselves spatially, historically and morally in relation to the Colombian state.

Jesús[17]

Some 20 kilometres west of Mocoa, as it leaves behind the rural settlement of Las Mesas and gradually plunges into the sharp folds of the *cordillera*, the road becomes narrower and much steeper. The transition is acutely felt from the cabin of Jesús's truck, a 1973 Ford F600 he has owned for more than 20 years. As it battles through the ascending hairpin curves or 'Lupas' – as this section of the road is suggestively known (Figure 4.2) – the engine's rhythm turns into a hoarse symphony composed of the ceaseless strained shifts of the gearbox: low range first gear, high range first gear, low range second gear, shift down again to high first and so on. This monotonous sequence is broken only by the sporadic approach of another vehicle coming down the road, when smaller vehicles yield to larger ones and must patiently wait for the other to pass. This tacit 'size law' of the road – single-lane for the most part – is not always followed and often leads to disputes and quarrels lasting until one of the drivers reluctantly gives up and undertakes the hazardous task of reversing.

I recounted to Jesús my personal experience months before while travelling from Sibundoy to Mocoa in a 'mass grave' – as buses making this route are infamously known – when we spent about half an hour face to face with a lorry whose driver refused to move despite the angry plea from the bus passengers. Jesús, however, who tirelessly defends truck drivers as the main victims of the *trampoline of death*, seems to absolve his colleague of any guilt in the dispute and instead evokes a time, back in the 1990s, when *traquetos* (drug traffickers) driving their brand new 4x4s often pointed their guns at them to make it clear who had the right

Figure 4.2 *Las Lupas* (2010).

of way. His story is followed by a talk on the misfortunes of his profession that goes on for a while until he makes a sudden stop in front of a small shrine on the side of the road. The stop is short but long enough for Jesús to light up a candle and urinate while Chepe, his long-time assistant, makes a quick check of the truck's wheels and brakes.

The shrine, housing the 'Virgen del Carmen' (patron saint of drivers), constitutes a ritual spot for road users, who regularly stop by and offer a prayer or a candle in exchange for a safe trip or as a way of paying tribute to those who have lost their lives on the road. Along with the other saints found along the road (there are at least five between Mocoa and San Francisco) this one has its own story. It was originally located next to the bridge over the Pepino River, 15 kilometres down the road, but was removed in 1977 after miraculously surviving a legendary flood that swept the bridge away. Initially relocated a few kilometres up the road, a few years ago it was vandalized – Jesús's suspicions fall on a 'satanic or evangelic sect' – so it was replaced anew and caged into a solid fenced chapel on Km 121, where it remains to this day.

Just a few hundred metres from the 'Virgen' lies a small altar commemorating one of the largest accidents of recent times, on 22nd July 2008, when a bus from Cootransmayo – Putumayo's main bus company – rolled off a 100 metre cliff killing 23 people. This tragic event bears a special significance, as it occurred two months after President Alvaro Uribe had publicly refused to support the construction of a bypass road to replace the *trampoline of death*.

* * *

As we resume the slow march across the Lupas, Chepe, a taciturn man except when it comes to road tragedies, brings up some of the lurid details of the Cootransmayo accident and evokes others signalled by the crosses and memorial plaques scattered along the way. There are many memorials that come in different shapes, colours, sizes and ages. Some look fresh and cared for while others are visibly eroded by the passage of time. Many, says Chepe, have been swept away by landslides or run over by lorries and buses. 'This road has too many deaths', he says while recalling one of its many names: 'The longest cemetery in the world'. And as many as there are unnatural deaths, there are stories of apparitions and ghostly characters inhabiting the road: 'El Gritón' (the screamer) on Km 110 and 112, 'the ghost car' between Km 115 and 125 and 102–103, 'the elf' around Fear Creek (Km 123) and 'the Indian' on Km 95, among others.

Jesús's and Chepe's vivid descriptions of the *trampoline of death*'s intricate landscape of the dead cannot but evoke the road's tortuous geography faced daily by its living users: ceaseless serpentine curves, debris from past accidents, long stretches covered by dense fog, lines of vehicles patiently waiting for bulldozers to remove fresh landslides. This exceptional 'roadscape' has been the subject of numerous terminal diagnoses ever since the time of its construction. In 1942, for instance, one of the road engineers noted that the Lupas section required 'a separate chapter' and gloomily observed that the 'forced' layout was 'inexplicable in every sense', the specifications 'couldn't be worse', and the drainage problems were 'fairly serious'. Nevertheless, he concluded that since the works on this section were too advanced, the only choice was to keep the existing route.[18] Almost 30 years later, a report on the road's landslides stated that in this same section: 'the use of explosives in its construction, the large masses of fractured rocks, the abundant vegetation, the violent nature of the region's winter, and the construction in loops, do not allow any measures to avoid landslides'.[19] The picture attached to the report (Figure 4.3) looks as desolate as its dismal verdict.

These textual and graphical descriptions of the road contrast with the official version of accidents on the road. Shortly before travelling with Jesús, I had spent some days at the Instituto Nacional de Vías [National Institute of Roads, Invías] reviewing the thick and dusty folders containing historical records of vehicle accidents.[20] Although records are only

Figure 4.3 Landslides across the Lupas, 1969.
Source: AGN, MOP, Box 6, folder 35, n.f.

kept for the past decade – older records had been lost for years and no one at Invías seemed to have a clue about this loss – the 121 accidents reported during that period along the 65 kilometres of the *trampoline of death*[21] constitute a large enough sample to get a sense of how accidents are reported by the Invías officers. The most prevailing 'possible causes' of accidents are listed under the following set categories: 'ineptitude in driving' (22 cases), 'driver's recklessness' (14), 'driver's drunkenness' (13), 'speeding' (13), 'sleeping' (9), 'fog, rain, smoke' (9), 'steering or mechanical fails' (8) or different combinations of the above. A few are attributed to road-related issues such as 'lack of visibility' (7), 'landslides' (5), 'road damages' (3), and 'wet surface' (1), while only one is exclusively blamed on the 'geometry of the road' and contains the following observation; 'highly dangerous section for different reasons: abrupt terrain, single-lane road, construction of bypass road urged'.

More detailed and imaginative descriptions can be found under the section on 'observations', such as this one from a truck at Km 78:

The cause of the accident was that the [truck's] gearbox slipped into neutral, crashing head-on with another truck; then it moves backwards 33 metres at great speed, turns sharply and falls off the cliff some 150 metres. Before falling, the driver's assistant jumps off the truck saving himself from perishing.

The accident is listed under the category 'imprudent reverse'. Another at Km 116, classified under 'ineptitude in driving', reads:

When approaching the curve, the driver pulls over too much to make way for a bus coming in the opposite direction and gets off the track, leaving a few seconds for the driver and his two companions to jump off the vehicle before it collapses.

And there are many more where, despite obvious contradictions, agency is variously attributed to human ineptitude, faulty gearboxes, brakes and steering wheels but very rarely to the road itself, whose 'inexplicable' geometry and 'fairly serious' problems appear to have become just another 'natural accident' within the landscape. Amid this normalized geography of official discourse there is an altered road warning sign at Km 95 (Figure 4.4) that ironically reminds the traveller of the pervasive presence of the road's other, moral geography. In the original sign ('Danger: zone of landslides') a letter E has been relocated so it now reads and translates as: 'Danger: zone of [falling down]'.

* * *

Figure 4.4   'PELIGRO. ZONA D DERRUMBESE'. San Francisco-Mocoa road, Km 95 (2010).

Jesús, clinging to the steering wheel of his truck (Figure 4.5) and skillfully negotiating the tortuous Lupas, looks much less concerned with the dead than the living. After all, he is a long time survivor of the road. He started in 1965 as trucker's assistant and became a 'patented driver' (holding a driving licence) in 1968. Ever since, he proudly declares, 'if I haven't been in the shop I've been on the road'. For the most part, he has covered the route from Puerto Asís and Guamuez to Pasto transporting all kinds of goods back and forth, from timber and livestock to construction materials, veterinary drugs, animal feed, gasoline, grain and all kinds of agricultural products. He has travelled the *trampoline of death* a couple thousands of times – between one and three times a week depending on demand – during the day and at night. He has lost a good number of friends to the road – he estimates between 60 and 70 – while many others have long since retired or have 'died naturally', and according to him only a few from the 'old generation' remain in the business. He has on numerous occasions escaped death, and yet watching his careful and skilled driving one could hardly conceive that these near-misses were due to 'ineptitude' or 'recklessness'. He acknowledges that in the past alcohol used to be a common issue among truckers and bus drivers, although he argues that there was a time when the road was so hazardous – he says the road nowadays 'looks like a highway' when compared with the past – that sometimes a 'drink of *aguardiente*' was essential to gather the nerves to drive through certain stretches.

Figure 4.5   Jesús behind the wheel of his truck (2010).

After all those years on the *trampoline of death*, Jesús's knowledge of the road has become corporeal: he recognizes and feels every pothole, every alteration of the road surface, no matter how imperceptible, every deficiency in its routine maintenance. And he resents it all. He insistently asks me to take note of every flaw he points out and to deliver them to 'some politician' in Bogotá, all the while complaining about his numerous pending or unanswered claims to Invías. Yet his indignation is not restricted to the neglect of the road and he enumerates five other 'factors' that have historically affected his profession: the police, the army, the guerrillas, the paramilitary and the robbers. These elements together comprise another geography, one that is highly dynamic and changing in time and space, complete with its own landmarks and 'hot spots': Los Guayabos' (Km 129), 'Curva del Km 126', 'Curva de los atracos' (Km 90) and 'Chorlaví' (Km 74), are famed for robberies; 'Curva de Ignacio' (Km 128) – named after a truck driver assassinated there – and 'Curva del Zambi' (Km 123), were guerrilla checkpoints back in the 1990s and early 2000s; and so forth.

Jesús and Chepe have been victims of muggings several times, although they recall the 1990s as the worst decade. Between 1995 and 2000 civil unrest spread across the region as a response to the government's *coca* eradication policies, a crop that at the time represented the main livelihood for a large part of the rural population. The conflict between *cocaleros* (*coca* growers), guerrillas and government brought to the surface the long-standing problems of the region, among which the lack of roads figured prominently and involved numerous strikes and mobilizations.

The FARC guerrillas set up road blocks and often banned the traffic of vehicles under the threat of burning them. When allowed through, road users were regularly charged expensive tolls, sometimes at various points on the road. According to Jesús, the fee between Puerto Asís and Pasto could oscillate between 80 and 200 thousand pesos (US$30 and 70 dollars) depending on the load value and number of tolls. When both the guerrilla and the army were absent, the road turned into a 'lawless territory' and assaults were everyday occurrences. Jesús and Chepe recall when they were assaulted seven times in just one day. On the last occasion, approaching San Francisco around 9:00 p.m., they were stripped of the last 6,000 pesos they had left for dinner. In another incident not long ago, recalls Jesús, he got stabbed for resisting an assault.

Even today, and despite the increasing presence of police and army on the road, waves of robbery are frequent. Yet on the *trampoline of death* even the perpetrators sometimes become victims. The Invías' records contain the following report of an assault on Km 116:

> The truck was heading from Mocoa to Pasto. According to the driver, two assailants, hooded, waylaid him and asked for his money. One of them tried to drive the truck but instead of driving forward drove in reverse rolling off the cliff.

This incident was classified as 'ineptitude in driving; attempted assault with a deadly weapon'.

* * *

Six kilometres ahead of the Virgen and nearly 400 metres above is the police checkpoint of 'El Mirador' [The Viewpoint], another 'rite of passage' of the road, noticeable from a distance by its towering telephone antennas and a tall statue of a Christ looking compassionately towards the Putumayo lowlands. The magnificent view of the Putumayo *piedemonte* and lowlands offers a stark contrast to the bleak landscape of the place: an old trench covered with wet camouflaged clothes, a long cement barrack, an improvised radio office, the remaining ruins from an old station shattered by guerrillas back in the 1990s, and two small 'casetas' or shops where travellers gather and wait for the routine police checks.

Jesús can't hide his disgust when having to confront the questions coming from outside the driver's window: 'Where are you coming from? Where are you heading? What kind of load are you carrying? Are you carrying any weapon sir? Can I see your IDs and the vehicle's documents please?'

Once the impassive officer takes the documents and walks away, we join a group of bus passengers crowded in one of the casetas and order coffee while Chepe makes another routine check of the truck. From there we can listen to the customary verification procedure taking place in the adjacent building: 'This is Alfa 3 over. Received Alfa 3. To verify a subject over. Roger Alfa 3. ID 8-0-4-0-7-7-4-2 over. Copied Alfa 1. (30 seconds). This is Alfa 1 over. Roger Alfa 1. Subject clean over. Roger Alfa 1 over and out'. Five minutes later the same officer returns with the documents and indifferently hands them over to Jesús with the words 'you are free to go'.

It takes Jesús's truck a good 20 minutes to overcome the final two kilometres of the Lupas and reach 'Filo de Hambre' [Edge of Hunger], an abandoned road station at Km 113 that now serves as temporary lodging for army soldiers patrolling the area. From there and for the next 20 kilometres the road becomes flatter, relaxing the truck's engine and thus attenuating the cabin's vibration. Jesús, still upset about the ten minutes we lost at *El Mirador*, begins a long monologue about his old distrust and revulsion towards the army and the police. 'We are told the police is our best friend but it's our worst enemy. They seem to forget transport brings life to the region' he states while complaining about the frequent mistreatment he and his colleagues have to endure daily: lengthy checks, sometimes numerous times a day, excessive tickets and constant requests for bribes. Yet what Jesús resents most is that despite being one of the oldest and most frequent users of the road he is treated invariably as an 'unknown'. This he considers a personal affront.

His relationship with the army is more complex. Although he acknowledges things have improved slightly over time (soldiers are now encouraged to greet travellers with a thumbs up sign), he thinks that until recent years they were the most serious threat on the road. He especially recalls the late 1970s and the 1980s, a period when guerrilla groups began to spread across the region and civilians ended up being trapped in the conflict with the army. Lorry drivers, constantly moving around the territory, were frequent victims of harsh interrogations and inspections. 'Those were awful inspections, we were ordered to take our shoes and pants off even when it was raining' recalls Jesús, while Chepe comments by way of clarification that 'at that time human rights didn't even exist here'. This humiliating situation, however, led to the most remarkable experience in Jesús's career. In the midst of the constant abuses by the military, he, along with many of his colleagues – Jesús estimates they were about a hundred – joined the M-19, a guerrilla movement that came to have significant political and military influence in the region in the 1970s and 1980s. Their role was mostly to provide information about military activity in the area but sometimes also included transporting firearms and other goods. In exchange, apart from training courses on guerrilla tactics and political indoctrination, Jesús proudly affirms, 'we got what we had historically been denied by the state: respect and value for our work'.

Although the M-19 demobilized in the late 1980s,[22] it left a deep mark on Jesús. He refers to this experience as a 'university' for those who, like him, 'were too weak to fight a heavily armed government'. Ever since his experiences with the M-19, he has aligned with leftist political parties and proudly defines himself as a 'fighter' and 'watchman' of the road, gathering signatures, grievances and denunciations from friends and colleagues. Yet he regards the nature of his job as the main obstacle to political action. 'We are a very disunited family' he says, pointing to the fact that he and his colleagues spend most of the time on the job and it is very difficult to meet beyond the brief encounters on the road. Besides, he argues, for independent drivers like him, making a living is so hard at present that they can hardly stop working. And as if this needed proof, Jesús cites by memory the poor earnings of an ordinary working day: the average freight of a trip from Mocoa to Pasto he estimates at around 600,000 pesos (10 tons at 60,000 pesos each). From this amount he subtracts the unloading fee (80,000 pesos), fuel (230,000 pesos), accommodation and food for two (100,000 pesos), and Chepe's wage of 40,000 pesos. Left with about 150,000 pesos (US$50), Jesús still has other costs such as taxes, insurance, union fees and maintenance of the truck: a tyre, for example, costs around 1,000,000 pesos and Jesús's truck has six; the average life of a tyre is about nine months but on the *trampoline of death* they burst more frequently. To make things worse, since the times of the *coca* boom, the region's agricultural production has greatly diminished, meaning that today his truck goes mostly empty on its journey to Pasto.

* * *

We have been on the road for four hours and travelled nearly 50 kilometres when we pass through the sector of 'Murallas' [Walls], famous for the single worst tragedy in the history of the *trampoline of death*. On the side of the road there is a small memorial shrine crowded with crosses and commemorative plaques, while others can be found fixed in the long stone wall along the road. All are inscribed with the date of 19th July 1991, when a huge landslide buried several vehicles killing dozens of people (there is no official number but estimates vary from 60 to more than 100) many of whom were never recovered. A long stretch of the road was swept away, leaving most of the Putumayo isolated for several weeks. A press note from *El Tiempo*, the country's largest newspaper, reported the event adding the following historical overview of the road:

> The entire road to the Putumayo is horrifying. The drivers that cross it daily call it the road of death. It was built during the conflict with Peru, way back in 1933. It looks more like a *trocha*. Moreover, travellers are constantly threatened by the guerrilla (*El Tiempo*, 21st July 1991).

As we leave Murallas behind and begin the last ascent of 'Portachuelo' – the mountain ridge that separates Mocoa from the Sibundoy Valley – Chepe relates some of the gory details of the tragedy: dismembered corpses and limbs covered in mud and scattered all over the place, bodies washed away by the avalanche and found days later kilometres away, bodies found intact but asphyxiated inside their vehicles, improvised mass graves to bury the 'NNs' (unknown or unclaimed). Jesús, less graphic in detail, mentions that he lost more than ten colleagues in the tragedy. He also says he doesn't like to stop there (he thinks the place is 'too heavy') but lately every time he passes by he thinks it's about time to retire from the job. At 66, he feels he has become 'nervous' and too tired of fighting with the Invías, the police and the road.

I ask him if with the *Variante* (the future bypass road from San Francisco to Mocoa) he might think about remaining in the business. 'The Variante is a great thing for Putumayo, but then even the "cocineras" (female cooks) from Mocoa's market will be able to drive a truck' he says, explaining how a 'modern' road will inevitably lead to the devaluation of his 'know-how', for better or worse linked to the *trampoline of death*. So his life as a truck driver, he believes, will be over with the demise of the old road.

* * *

Along with Jesús, there are others for whom the long awaited replacement of the *trampoline of death* is a point of concern. 'Malacate' (Km 110), a *colono* from San Francisco who used to make a living by rescuing cars ('If you tie a cow to a tree she'll be hung next day', he told me once in explaining the impossibility of setting up a farm along the road), a role that is now less common since, as he explains: 'the road is wider, people drink less and the trucks are now insured'. And he goes on to claim the government projects offered to the few inhabitants of the road are useless: 'with the new road not even the birds will visit us here, so what the hell am I supposed to do with a trout pond here?' he asks indignantly. His plan, he says, is to sue the government to get compensation for the land plot he owns and use the money to set up a business in his hometown. María Cárdenas, restaurant owner (Km 106), is sceptical about the plan to turn the *trampoline of death* into a tourist attraction. 'If the government is not able to maintain the road as it is now', she asks, 'how is it going to do it in the future? We are going to be abandoned here, die of hunger'. Laureano Cusi (Km 129), a road maintenance worker for 24 years, fears for his pension: he is 47 and still has 13 years to reach the retirement age; yet, he is pessimistic about the *Variante* and doubts it will be ready before his time arrives.

Jesús is sceptical about the future road too, whose slow progress he attributes to the opposition to the project on the part of the indigenous communities. 'All they want is money from the government', he says visibly irritated, and then goes on with an explanation I had recurrently heard amongst non-indigenous Putumayenses and which mirrored the pervasive nature of the state's old civilization rhetoric:

> Because they are not productive people, they are lazy and want the government to feed them, to provide them with health, with subsidies, all that. And they have to thank the government and the whites for what they have been given, education and all, otherwise they would still continue to be savages.

Since this issue seems beyond dispute for Jesús, I change the topic and ask him what his plans are for retirement after almost five decades on the road. He talks about buying a piece of land near Mocoa to set up a small agricultural farm, an unfulfilled dream from his youth, and conjectures at length about the potential crops and the profits he could make on them. Yet, as if waking up from the dream, he says that after so many years on the road he finds it hard to imagine himself 'anchored'. Thus, he will maybe replace his old Ford for a dump truck when his truck driver license expires a few years from now, when he turns 70; or perhaps he will buy a taxi to carry passengers along the new road from Mocoa to San Francisco; or who knows.

As Jesús continues speculating about his future plans we suddenly come across a police checkpoint on Km 75, just when we are about to conclude our journey through the *trampoline of death*. This time he is asked to step out of the truck and take the vehicle's documents with him. From the truck's cabin we can't hear his conversation with the officer, but from its high tone we can suspect something is going wrong. After five long minutes Jesús is back and without a word starts the engine and drives off.

'So?' asks Chepe eventually. 'The door's registration seal' answers Jesús with a tired expression and explains to me that it was recently removed when he sent the truck to be painted. 'And how much?' replies Chepe. 'Fifty thousand', comes the reply.

Franco

I first knew about Franco in February 2010. Reading an online opinion column from a local newspaper, I came across a comment with the title (in caps) 'DO NOT MAKE POLITICS WITH THE PUTUMAYO ROADS' and signed 'Franco Romo Lucero: Overseer Pasto-Mocoa road'. The comment, longer than the article itself, was directed against Orlando Guerra, a congressman and candidate from the Putumayo to the National Parliament whose re-election campaign motto was 'the parliamentarian of the roads for development'. Franco emphatically (and ardently) accused Guerra of claiming that progress on road development in the region – specifically the recently approved project to build the bypass road aimed at replacing the *trampoline of death* – resulted from his personal influence. 'In my 35 years working as an agricultural engineer in the Putumayo', said the commentary, 'I never heard that Dr Guerra had participated in any meeting, committee, forum, debate, etc. concerning the Putumayo roads'. Thus, he demanded Guerra to 'respect' the enduring popular struggle for 'decent roads', which he vindicated in the following terms: 'Since 1955, it was the leaders from Mocoa, together with 45 leaders and 23 ex-mayors from the towns of Santiago, Colón, Sibundoy and San Francisco, the ones that joined together and began to advocate for better roads in the Putumayo'.

Franco's aversion towards Guerra's campaign, as I would later learn, was shared by many. Every time I brought up the question of his campaign (unsuccessful this time) I heard the words 'unfair', 'abusive' and 'unjust' to describe the politician's self-proclaimed role in the pro-motion of roads. Those responses, moreover, revealed how sensitive the issue of roads is in the region and in particular within the local history of popular struggle. This struggle, like Jesús's 'political topography' of the *trampoline of death*, has a manifest moral content, in this case inex-orably connected to the ways in which people invoke the past to make claims about the present and future. Put another way, history appears here, as notes Rappaport, as 'a question of power in the present, and not of detached reflection upon the past. It can serve to maintain power, or can become a vehicle of empowerment' (Rappaport 1998, p.16). It was precisely by appreciating this political value of history in the struggle for roads, that I would eventually learn why a character like Franco Romo is so familiar in the Putumayo.

Franco is a native of Colón, where his parents, peasants from Nariño, arrived in the 1930s. At 73, he has had a long and varied career that includes being a school teacher, Secretary of Planning, of Development, and of Agriculture in the Putumayo, as well as leadership positions in numerous civic and community organizations (posts that, he emphasizes, he achieved 'thanks to personal merit and not to any political favor'). During the past three decades, however, he has been involved in the struggles around the Pasto-Mocoa road, organizing various pro-road committees, debating on the radio, promoting strikes, sending complaints to politicians in the Putumayo, Nariño and Bogotá, writing in the local press and, more recently, engaging in intense debates on internet forums.

Franco soon became a point of reference in many conversations I had about the past and present history of the road. He was often referred to me as someone 'que ha peleado mucho por lasvías' [an enduring fighter for roads] and a very knowledgeable person of the history of struggle. Following a few emails and phone conversations, I eventually travelled to Pasto, where Franco currently lives, to meet him. We spent a morning at his house, Franco giving me a long autobiographical history of the road, covering topics such as his admiration for the Capuchins' tenacious campaigns for the Pasto-Puerto Asís road (of which he heard constantly during the years he attended primary school at the Mission's seminar of Sibundoy), the 'villainous engineers' responsible for the existence of the *trampoline of death* (his father had been a food supplier during the construction of the road and had given him the names), the worst tragedies (his brother lost a truck with its driver and two passengers in the Murallas accident), the early times of the popular struggle (which he joined in the 1970s), the countless unfulfilled promises of a new road and in-vain lobby trips to Bogotá (where he was often reproached 'for coming with the same old sermon'), and his most recent quarrels with politicians.

I learnt that his latest dispute was not only regarding the *trampoline of death*, but also around another section of the road which the government had recently opted to pave, despite engineering studies recommending its replacement due to design problems and environmental impacts. Eager to show short-term results, Guerra and other local politicians had initiated a smear campaign against Franco, calling him 'public enemy of roads' and 'officially' declaring him *persona non grata* in the Putumayo, to the point that for some time he was threatened with lynching if he showed up in Sibundoy. (Weeks later in Mocoa an Invías officer would tell me 'off the record' that the ongoing resurfacing work was indeed a huge mistake, and during my trips with Jesús we saw some overturned trucks along that section, which he angrily blamed on the road layout).

Franco's account of this conflict allowed him to make a broader point about the historical lack of leadership and 'short-sightedness' of most politicians from Nariño and Putumayo which, together with the 'negligence of the state', he considered to be the main cause for the precarious and backward state of roads in the region. Yet this argument, which I heard many times, allowed him to highlight again the long tradition of popular leaders and struggles, as well as his own legacy in this process. He jealously kept an archive of memos, letters and written complains dating from the late 1940s to the present, some of his authorship and others that he had gathered or inherited from past leaders. At the end of our meeting, I asked him for a copy of this archive, which he generously agreed on condition that it was photocopied at 'un lugar de confianza' [a place he knew and trusted], to which he personally escorted me.

Days later, examining Franco's archive, I realized why he treasured these documents so much. The archive was small, no more than 40 pages in total, yet together they comprised a comprehensive genealogy of the struggles for the road, relating different characters, landmark dates, laws, press articles and minutes of community meetings. Many were heavily marked, and I would later notice Franco habitually resorted to them in his own writings, often reproducing them *in extenso*, revealing the value of the past both as testimony and factual evidence supporting and making sense of present struggles. One of the documents, entitled 'A faithful testimony of a fight for the social and economic demands of the Intendancy of the Putumayo', particularly struck me for both the perseverance of these frontier leaders and the state's bureaucratic apathy and inertia. The three-page typescript document, authored by Isaías Rosero, another renowned leader from San Francisco, compiles several chronological responses to petitions about the replacement of the *trampoline of death* sent by him to various public offices in Bogotá from 1963 to 1979. Some of them include:

No.1897. 20th December 1963. Regarding construction road San Francisco-Mocoa. Road layout studies ready. Waiting to know national budget for next year to begin construction with allocated resources for such road. Sincerely, Castillo, chief engineer, Ministry of Public Works.

No.CP-858. 13th June 1967. Dear Sir: Regarding the petition you make in your letter, I inform you that a programme for the Popular Integration of the Putumayo is to be launched, and within that programme one of the projects which will be given priority is precisely the San Francisco-Mocoa road. Sincerely, Emilio Urrea Delgado, Presidential Counsellor.

No.28853. 10th July 1975. Dear Sir: I acknowledge receipt of your communication from 3rd July 1975, in which you make reference to your previous one of 30th September 1974. I gladly inform you that this Ministry, aware of that rich region of the Putumayo, has determined to give a prompt solution to the San Francisco-Mocoa road. Sincerely, Humberto Salcedo Collante, Minister of Public Works.

No.T:00415. 7th July 1977. Dear Sir: Regarding your communication from 28th May, where you raise the need for the construction of an alternate road from San Francisco to Mocoa. In this respect I inform you that this Department shares your concern, and that within the current budgetary constraints your request will be taken into account in the elaboration of the 1978 budget. Sincerely, John Naranjo. Director, National Planning Department.

No. 008. 9th January 1979. Dear Don Isaías. I have read with interest your letter sent late last year, together with your article 'An absurd and dangerous road' published in the newspaper *El Putumayo*, and your letters to the President, the Minister of Public Works, and the pamphlet in defence of the alternate road. All these documents represent a valuable and convincing defence for a solution to the actual road. I agree that the current road, especially after the tragedy of 6th November, constitutes a precarious solution and should be replaced ... However, the solution that you mention, the alternate road between San Francisco and Mocoa, of which 12 km were already built, had to be suspended due to topographic and geologic difficulties. Nature, having gifted the Putumayo with much of the country's best lands such as the Sibundoy and Guamués valleys [sic], has also put formidable obstacles disconnecting Putumayenses. This is why we should study these projects patiently and imaginatively, to resolve obstacles in appearance insurmountable [sic] ... The Ministry of Public Works has changed its politics substantially and demonstrated a great concern for the Putumayo. Kind regards, Gustavo Svenson Cervera, Director, Administrative Department for Commissaries and Intendancies.

Rosero's endless pleas never turned into more than vague promises of bureaucrats from Bogotá. However, they were later picked up and reproduced by Franco in his 'Historical chronology of the organisation and struggle', a chapter of a book he published in 1990 entitled *Carreteras Variantes* [*Alternative Roads*], an original copy of which he keeps in his archive.

The back cover of the book (Figure 4.6) shows a sketch of Franco in a combative pose, wearing a sash with the title 'Las Variantes', alluding to the road sections between San Francisco and Mocoa and Santiago and El Encano, and surrounded by the slogans 'we will fight' and 'we will defeat'. The illustration not only introduces the author but is highly suggestive of the book, largely a compendium of writings (historical,

Figure 4.6   Back cover of Franco's book.
Source: Romo (1990).

political, autobiographical) and documents (press articles, road proposals, legal complaints, public and private ads advocating for the road),
aimed at both documenting the struggles and justifying the need to continue fighting (the 'exclusive purpose' of the book, as stated by the
author, the then President of the Committee 'Pro-*variante* San Francisco-
Mocoa', was to make a call to send a commission to Bogotá in order to
'remind the President of his promise' and request a budget for the road).

The entire image, on the other hand, showing Franco emerging from and
dominating a map of the Sibundoy Valley (traversed by the drainage works
in which Franco proudly participated as engineer in the 1970s), also evokes
another story: that of the twentieth-century's dramatic transformation of

the valley initiated by the Capuchin mission, and the hegemonic ideals it sought to implant and appear here embodied in the figure of the *colono* and the built environment. The first chapter of the book ('The conquest of the Amazon through a road'), for instance, is an enthusiastic eulogy to the Capuchins' enduring fight for the Pasto-Puerto Asís road, which the author finds inspiring for the present and future struggles, thus commenting that: 'Now, in 1989, we should fill ourselves with courage, follow their example and undertake a struggle leading us not to civilization, but to the development of our Amazon resources' (Romo 1990, p.5). This aspect is of crucial importance, as it reveals how in the act of invoking the past to make sense of the present, some stories and characters are reified (the missionaries, the past *colono* leaders and the struggles for the road) while others (the Indigenous dispossession of land and labour associated with the road) remain silent or are simply excluded from the picture.

Despite those silences and erasures, to which I will return later, Franco's history exposes the road's moral geography, a geography that no inhabitant of the Putumayo can escape and which he consciously underscores to show the frontier's sense of neglect and abandonment. Although there are several parts of the book where this sense comes to the fore, I will refer to two examples where it becomes especially clear. The first one concerns the burdensome experience of travelling across the *trampoline of death*, which Franco accentuates by contrasting it with the journey along another road, and describes in the following terms:

> Those who travel between Pasto and Mocoa have to pay $9,40 per km, bearing all the discomforts, shaking, frights and risks of being buried into the cliff; in comparison with the $5,71 per km paid by those travelling between Pasto and Cali in Pullman buses, five stars with TV and a stewardess on board. We know that the more comfortable and faster the trip is from one destination to another, the more expensive the ticket; which doesn't happen to us Nariñenses and Putumayenses for whom, the more gruelling, riskier and slower the trip, the more we are charged. So, why is it this way? Who can explain this dilemma? Will we have to resign our children's fate to this bleak future? What do you think? Let us know (Romo 1990, p.15).

The second has to do with the 'martyrs' of the *trampoline of death* during its '50 years of existence', which the author gathers from 'fragmentary evidence'. According to Franco, the road 'has claimed more than 2,700 lives among workers, drivers and passengers, and left some 3,900 invalids, together with 5,700 million pesos in losses of machinery, vehicles, merchandise and other goods that have been buried in the cliffs of that gloomy way' (Romo 1990, p.20). The significance of these numbers, however, does not seem to lie in its magnitude or even accuracy (for no source is mentioned), but in how those victims represent, in Franco's words: 'the weapon that, joined to the democratic struggle, will lead us soon to the conquest of this long-awaited alternate road' (Romo 1990, p.21).

***

Twenty years after the publication of Franco's book, not one kilometre of the bypass from San Francisco to Mocoa had been built. News and rumours sporadically announced the beginning of the works 'next week' or 'next month'. And though some looked expectant and optimistic with every new rumour, many, especially the old, sounded sceptical and often assured the future road was one of those things they would never see come true. But Franco, despite the many decades of struggles and failed promises, looked far from defeated. At the time I met him, he had recently organized a 'veeduría ciudadana' [citizen oversight committee] for the *Variante* project, and told me he was preparing an upgraded version of his book. If things go as announced by the government, he thinks, he will be around 80 by the time the 45-kilometre new road is concluded.

Guillermo

Guillermo was born in 1938, into a family of *colonos* from Nariño who settled in Mocoa in the early 1900s and had prospered during the rubber boom. When he was around 14, his mother, portrayed by Guillermo as a driven and strong-minded woman, encouraged him to continue his education outside the Putumayo, where schools were at the time controlled by the Capuchin Mission. Through the Bishop of Sibundoy, he got a scholarship for the Colegio Nacional de Cartago, a school in one of country's central departments, some 700 kilometres from Mocoa. It was late September when the Bishop gave him the news in Sibundoy, and ordered him to turn up at his new school by the first week of October. He took a truck to Mocoa, where his mother hurried to pack his bag and asked his elder brother Víctor to accompany him on the long journey.

The longest and most torturous part of the trip was the 148 kilometres from Mocoa to Pasto. In those days, it was usually made on top of lorries packed with plantain, timber and other products from the region, and could take from 12 hours to several days depending on the usual contingencies such as road accidents, landslides or breakdowns. From

Mocoa to San Francisco, where the entire road was single lane and rarely more than three metres wide, there were various 'pidavía' stations (checkpoints connected by crank phones through which operators controlled the up and down traffic). Waiting times at each station could vary from a few minutes to several hours, depending both on the mentioned eventualities and the time of the day or week.

Guillermo and Víctor travelled by night, when the traffic was usually lighter, and made it to Pasto safely and without any major delays. However, in Popayán, half way between Pasto and Cartago, they had to spend a few days, Víctor having to attend some personal business there. They reached their final destination in mid-October, just to find, to their great dismay, that the students' registration dates were already closed. For a few minutes, Víctor begged the school's secretary to no avail. Then, when things seemed totally hopeless, Guillermo took the floor and gave what he describes as his 'first public speech'. Almost 60 years later, from the terrace of Junín, the farmhouse built by his father and which Guillermo inhabits today, he tells me the story as if it had happened yesterday:

> I said to Miss Celina (the secretary): 'how is it possible that I come from Mocoa, in the Amazon jungle, son of a *colono*, and you reject me?'
>
> I remember there were three desks in the room and in the third one there was a man listening. At some point, he turned to her and said: 'Celina, tell the boy to come here'.
>
> And he said to me: 'I am the National Inspector of Public Schools. Who sent you?'
>
> –the Bishop of the Putumayo
>
> Then he looked at a map of Colombia hung on the wall and asked me: 'Where is the Putumayo?' How long did it take you to get from there to here?'
>
> I pointed out Mocoa on the map and said: 'ten days Sir'.
>
> He hesitated for a moment but then turned again towards Miss Celina and said: 'Miss, please register the boy at once'

I first heard about Guillermo (Figure 4.7) a few weeks after moving to Mocoa, through an engineer working at the local Invías office. Guillermo, he told me, had worked there as a field engineer for many years, until he retired in 2004, and was a sort of 'living archive', to the point that he was occasionally asked for information regarding past contracts, road works, technical studies and the like. However, I only came to understand this fully when I met him in person some days later.

Figure 4.7   Guillermo Guerrero. Junín (2010).

I had briefly mentioned on the phone that I was interested in knowing about the history of the *trampoline of death* from an engineer's
point of view. He addressed this issue during our first meeting at
Junín, in a two-hour photographic presentation depicting the road
kilometre by kilometre. Through Guillermo, I was first introduced to
the intricate geography of cracked road surfaces, unstable foundations, endless faulty layouts and hairpin bends, eroded slopes and
damaged roadside ditches and culverts 'patched like old trousers'. His
overall verdict of the *trampoline of death*, which he gave me at the end
of his long presentation, was however much less technical and much
more emotional. 'The design of the road, the curves, all that you saw',
he declared as if trying to make sense to himself of an event he could
not otherwise digest, 'mirrors the nature of Colombians. The
Colombian has a twisted nature, a treacherous nature. I could never
understand that madness'.

But 'making sense' of the *trampoline of death* or deconstructing his
lapidary judgement, as I would learn through the many afternoons spent
at Junín, was a very complex issue for Guillermo. Besides his very-
detailed knowledge of the road, he had a prodigious memory and pas-
sion for history that went far beyond the anatomy of underground
drainages and landslides. For him, as for Franco or Jesús, the past
revealed itself as a vital element to understand and give sense to the present struggles around the road or the exclusionary and violent space of

its potholes and police checkpoints. Yet, and unlike Franco and Jesús, for Guillermo history was not restricted to past memories or the historical event confined to the short or even mid-term. Rather, it was only through long time spans or the *long duration* that present time or events made sense to him. Therefore the *event* for Guillermo, as in Braudel, appeared as 'part of a unit of time much longer than its own duration', or as 'an indefinite chain of events and underlying realities, which seemingly cannot be separated from one another'; and *time*, consequently, 'slowed, often to immobility' (Braudel 1960, p.4, 7).

As a long time inhabitant of the Putumayo and descendant of early-twentieth-century *colonos*, *neglect* was a key term within Guillermo's speech, one that summarized the state's policies (or lack of them) in the Amazon. Its origins, however, Guillermo insistently repeated, long preceded the birth of the Colombian state and had to be traced back to the border treaties between Spain and Portugal. From *Tordesillas* back in the late-fifteenth century (when the kings of both empires first shared out the recently discovered territories outside Europe) to *Madrid* and *San Ildefonso* in the eighteenth century (through which Spain acknowledged the *de facto* expansion of Portugal in South America well beyond the originally agreed limit) *neglect*, cried Guillermo, was the word to describe the Spanish authorities' virtual absence in and lax attitude towards their vast Amazon dominions. 'And after the twentieth of July 1810, Colombians did nothing to change this state of affairs' he said indignantly, blaming the *criollo* founders of the Republic for having inherited their predecessors' long and 'unjustifiable' tradition of having an Amazon *tirada a la come-mierda* [thrown to shit].

It was against this 'background' that the figure of Rafael Reyes burst onto the scene. Guillermo introduced him quoting by memory the intro-ductory excerpt of the General's Memoirs chapter on the Putumayo: 'In Pasto the region that extends to the east was known only as far as Mocoa, and beyond there the populace, ignorant of geography, thought it was Portugal'. The phrase, opening the chronicle of Reyes' self-pro-claimed 'discoveries' and 'conquests' in the Putumayo during the 1870s, allowed Guillermo to signal a landmark event in the region's history. The event was Reyes himself, and its significance so decisive from Guillermo's point of view that he did not hesitate in claiming that 'the active presence of the state in this corner of the Amazon began with Reyes'; 'because looking back', he added, 'there was nothing'. This sig-nificance, moreover, was twofold. First, as an entrepreneur, the *physical* presence and displacements of Reyes in the Putumayo, through its *pára-mos*, *selvas* and rivers, together with the nationalistic tone impregnating his journeys, had left according to Guillermo an indelible *corporeal* trace of the state on its remote lands. 'As an individual', recalled Guillermo, 'Reyes *knew* Pasto, Mocoa, what is today Puerto Asís and Puerto Leguízamo, Iquitos, Manaus, Belém. And from there he *projected* all his activities to Rio de Janeiro, Lisbon, Paris, London, New York'. No matter that these activities were mostly carried out for self-profit; 'What really matters', he claimed, was that through them, 'the Putumayo first appeared on the map'. Even more significantly, as I noticed from Guillermo's exalted emphasis of the idea that Reyes 'projected' those activities *from* and not *to* the Putumayo, was that the frontier, at least momentarily, surfaced at the centre rather than the periphery of this same map.

Secondly, as President, stressed Guillermo, Reyes began to translate into practice the 'vision' he had developed decades before 'as citizen and dreamer'. 'What was Reyes' vision?' he asked leaving a solemn pause – as if anticipating its grandeur and redemptive power – 'Reyes' vision was to connect the whole Amazon region to the world by establishing a trade route from Tumaco in the Pacific to the mouth of the Amazon in the Atlantic'. The General's support for the Pasto-Mocoa road, which Guillermo was always eager to explain to the slightest detail through the map he kept from Miguel Triana's expedition (Chapter 2), could only be understood within this much broader and ambitious scheme. At the centre of this plan, speculated Guillermo, was Mocoa, a 'fact' mirrored in Reyes' measure of creating the Intendancy of Putumayo with this town as its capital. 'And it was Reyes', he asserted, 'who recognized for the first time in history the name of Mocoa', and went on criticizing those who argue that Mocoa is the oldest town of the Colombian Amazon and 'rivals in age with Bogotá'. For Guillermo, it was clear that the difference between 'exist' and 'not exist' lay not in age but in the gap between *recognition* and *neglect*.

So imposing was the image and legacy of Reyes for Guillermo that he could easily reconcile the failed existence of the Intendancy (it lasted one year), or even the fact that the General's interoceanic plan never materialized beyond the ambitious map he presented at the Pan-American Conference in Mexico. As for my references to the Putumayo Concession scandal, which directly involved Reyes (Chapter 1), he seemed to pay little attention to them and instead went back, just like Reyes did in his time, to recount again the General's nationalistic enterprises.

Dealing with the fact that the same 'legitimate founder' of Mocoa had established a Penal Colony there for his political prisoners was a much thornier issue for Guillermo. More than a century had passed since this event took place and yet he declared himself 'incapable' of reading the infamous presidential decree. For Guillermo's real trauma was rooted not in the isolated and short-time consequences of the legal act but its long-term effects. And here a third term, *resentfulness*, was added to his *long duration* historical vocabulary of the frontier. How it emerged had a straightforward explanation according to Guillermo: the Penal Colony had 'hit the heart' of a good number of Colombians including the prisoners themselves, their wives, relatives and friends, hence creating a 'negative image of Mocoa', which eventually turned into a 'national resentment' towards the place and the whole Putumayo. Reyes' implicit responsibility in this episode constituted an issue Guillermo admitted to but did not like to dwell on, limiting himself to saying that it was perhaps 'the desperate act of a statesman' yet something he, Guillermo Guerrero, 'could have never reconciled in mind and heart'.

Guillermo's somewhat schematic and simplified connection between the event of the Penal Colony and a supposed 'national' rancour towards his hometown, had nevertheless an inner logic. As with most Mocoanos, for him this event was fundamental to make sense of another event that left a long-lasting scar on the town and its people: the construction of the *trampoline of death*, and more specifically the fact that this road avoided the Putumayo's capital. Here Guillermo's statement was categorical: 'The road was built to punish Mocoa', he said emphatically, and proceeded to narrate his version of the story, the same I had heard on multiple occasions except for the exceptional attention paid to detail:

> In 1927, the government of Abadía Méndez orders the studies for the road Pasto-Puerto Asís. Mocoa was not mentioned in the study. The engineer responsible for the studies was Alfonso Paz, from Popayán, son of one of the prisoners of the Penal Colony. Then came the war with Peru, and the government of Olaya Herrera urgently orders the construction of the road 'como se pueda' (by any means). And who is put in charge of the road? The engineer Sebastián Ospina, nephew of another prisoner. And listen: both Alfonso and Sebastián said in public that as long as they were the managers, the road would never pass through Mocoa, because the road was a revenge against the town of Mocoa. And they explained. They said: 'because my father or my uncle was there, imprisoned by a barbarian named Reyes, the dictator Rafael Reyes'. And when the road eventually arrived to Pepino and the people of Mocoa made their first trip to see, they could corroborate that madness, that nonsense!

As if in its 'madness' and 'nonsense' the episode demanded evidence to prove it was real, Guillermo listed people, 'regrettably dead', who met the engineers and personally witnessed the moment they pronounced their cruel sentence. Then, he went on to tell of the heavy toll paid by Mocoa as a consequence of the engineers' revenge: of its abandonment and isolation for more than 60 years, until 1991, when the Mocoa-Pitalito road was opened (according to Guillermo a 'compensation measure' that was offered by the government since the late 1940); of its decline and widespread poverty that came when it was converted into a 'peripheral settlement within a peripheral region' – which he illustrated with his own family, once well-off and then reduced to 'self-subsistence'; and of its many decades of struggles and unfulfilled promises that took the town and its afflicted inhabitants back to their immemorial condition of *neglect*.

'Was the conduct of certain Colombians against the Putumayo and Mocoa justified?' asked Guillermo visibly irritated upon finishing his recount of the decades of misfortunes suffered on account of the evil engineers, 'No, but the heart of Colombians is like the *trampoline of death*, twisted and treacherous'. His rhetorical question seemed both intended to lay blame on a past individual crime and at the same time on a much more abstract condition and intangible offender. For, beyond the need to 'make sense of the nonsense' by seeking scapegoats and resuscitating witnesses, it was evident that the 'offence', still unsettled, was rooted in the long chain of events. His ensuing claim, both retrospective and prospective, was illuminating:

> The Colombian state has a social and economic debt with the Putumayo. Why is there such a debt? Because of its oil, its precious woods, for the fact of being a national territory, for the people that inhabit this territory. What do we need? We need a modern road between Mocoa and Pasto. What does a modern road mean? A road with a design speed between 40 and 60 kilometres per hour, so that we can travel from Mocoa to Pasto in 2 hours and not in 18 like in the past or 7 in the present.

Yet Guillermo's views on the significance of this 'modern road' extended way beyond the reduction of travel times between Mocoa and Pasto, and it constituted a subject he could talk about for hours. Broadly, this road meant *physical* 'connection' and 'integration', and here he lectured at length about the new road being the core segment of the IIRSA Tumaco-Belém interoceanic corridor, the scheme that would at last see Reyes' vision realized (he himself had his own parallel proposal linked to this project: the Amazon Road Network, a plan he had presented years ago at a national congress on transport infrastructure, and which basically contemplates a network of river and road routes linking Colombia, Brazil and Venezuela, and has Mocoa at its geographical centre). Most essentially, however, this road meant a long-awaited infrastructure that would finally 'redeem' both Mocoa and the Putumayo from its *long-lasting* peripheral and neglected condition. And here he asked again, as if hoping perhaps the question itself might bring the government to reason: 'for what sense does it make to Bogotá to keep having a Mocoa, a Putumayo, an Amazon, thrown to shit?'

Uneven frontiers

Guillermo and I could never entirely agree about Reyes' interoceanic scheme or its resuscitated version under the IIRSA Tumaco-Belém mega project. Where I saw in the project the continuation of certain political and economic logic, he tirelessly emphasized its countless benefits for Mocoa, the Putumayo and the Amazon region in general. When we sporadically came across my research, he would put forward arguments he deemed irrefutable, and often concluded our discordant dialogues with the same statement: 'In any case, don't forget to mention that roads are fundamental for development'.

With time, this periodic ritual of irreconcilable visions revealed to me how the act of making sense is deeply rooted not only in *how* we situate ourselves against events, but *where* we are situated in relation to them. Guillermo's perceptions, for instance, are those of a historian for whom events only make sense in connection to others, often located in very distant times and places. Franco's, meanwhile, are strongly attached to the genealogy of struggles and the ways in which he situates himself in the lineage of popular leaders. And Jesús's, in turn, are mediated by his emotional relationship with the road's deadly topography and daily conflicts with the police and the army. Still, all these perceptions are rooted in and emerge from a same spatio-temporal *situation* and they all express a similar sentiment: a sense of occupying a marginal position in the Colombian state or, more specifically, of being included into it through a series of exclusionary – historical, spatial, political and symbolic – acts.

This sense, as I tried to illustrate in the first part of the chapter, is related to the vital significance, both physical and imaginary, that roads have had in the life of the Putumayo, and the ways in which they materially and symbolically incarnate ideas such as 'civilization', 'progress', 'development' and so forth. As these ideas lie at the core of the narrative in which the project of the state is founded, the dramatic history of the road and the tragic *event* of the *trampoline of death* plainly revealed the obverse side of this same narrative: isolation, abandonment, confinement, backwardness, neglect. Although the everyday realities these terms convey seem to conjure an image of remoteness or even absence of the state, they ultimately render its presence more visible: through Jesús's abhorred Invías officers, police and army checkpoints; Franco's ignominious local politicians and archive of unsuccessful grievances to bureaucrats in Bogotá; and Guillermo's neglectful colonial authorities, insensitive nation's founders, and treacherous engineers.

There are two elements or dimensions present in the narratives of these characters, both of which are crucial within the relation of inclusive exclusion between frontier and state. The first element is about the way in which the three narratives speak of how Jesús, Franco, and Guillermo, in *consciously* situating themselves and others not merely as geographical and historical but essentially moral subjects, or alternatively, in apprehending history and space morally, make sense and call into question this relationship. Put differently, these are narratives that, though in different ways, express both the pervasiveness of this relationship in the lives of frontier peoples and the way in which they contest it.

The second element is more complex and has to do with how we can overcome the idea of the state and frontier as homogeneous and antagonistic realms. In this and the previous chapters, I have addressed this question by exploring the material and discursive practices through which these realms are constructed. I have repeatedly emphasized that what defines these practices is the relation of inclusive exclusion through which the frontier has been incorporated into the state's political and spatial order. In doing so, the different characters who have been introduced to reconstruct and interrogate the spatial history of the Amazon frontier, have been described as embodiments of this relation rather than personifications of the state or the frontier.

Some of those characters speak of how such a relationship was conceived, while others mirror the different ways in which it is experienced, reproduced and enacted. This is the case of Jesús, Franco and Guillermo, whose narratives reflect not only the physical and symbolic violence through which the Putumayo was constituted as a frontier space, but also how this violence is unevenly distributed across race and place within this very space. This is particularly evidenced in the stories and characters these narratives give prominence to (the Capuchin mission and the *colonos'* legacy in the past and present history of the Putumayo, Rafael Reyes' vision and projects aimed at unleashing the untapped riches of the Amazon, the Indian's 'debt' with the 'whites' for helping them overcome their 'savage' state), and those that they silence or render invisible (the violent dispossession of indigenous lands following the opening of the road, the continuing conflict between 'whites' and 'Indians', the many social and environmental conflicts surrounding the current road project).

These silences all stem from a broader hegemonic narrative through which the frontier was constituted, and whose origins were already discussed. In this sense, they can be better understood if we think of them not as intentional or conscious silences but silences that reflect how dominant narratives interpellate subjects in different ways. Following Trouillot (1995), these are silences that ultimately mirror the uneven power in the process of production and narration of history, and thus have to be traced back to the origins of this process. But these silences are also crucial within the spatial history of the frontier, for they reveal how violent practices and conflicts present in this spatial history manifest or express themselves in highly differentiated ways among individuals, communities and places.

Notes

1 On the notion of 'moral geography' see, particularly, Thomas (2002); see also Harvey and Knox (2008) and Thevenot (2002).

2 'Decreto 1652, 30th September 1926. Por el cual se organizan los trabajos de trazado y construcción de la carretera de Pasto a La Cocha', 'Decreto 1245, July 29, 1927', AGN, MOP, T.1406, fols.41, 167; Jeremías Bucheli to Minister of Public Works, 8th May 1928, AGN, MOP, T.1406, fols.373-375.

3 'Informe de los trabajos verificados en la carretera nacional Pasto-La Cocha', 8th August 1928, AGN, MOP, T.1406, fols.419-421v.

4 Presidencia Asamblea to Consejo Nacional vías de comunicación, 5th March 1929, AGN, MOP, T.1406, fol. 606.

5 Rafael Agudelo to Minister of Public Works, 11th December 1932, AGN, MOP, Vol.3273, fol.199.

6 'Decreto 1549, 26th September 1932. Por el cual se dispone la ejecución de algunos trabajos', AGN, MOP, Vol.3273, fols.1–3.

7 'Decreto 1581, 28th September 1932. Por el cual se fija el plan de trabajo que debe adoptarse en algunas vías', AGN, MOP, Vol.0318, fols.8–11; AGN, MOP, Vol.3278, fol.36.

8 Agudelo to Minister of Public Works, 11th December 1932, AGN, MOP, Vol.3273, fols.199–214.

9 Agudelo to Minister of Public Works, 1st January 1932, AGN, MOP, Vol.3545, fols.224–231.

10 APPC, Missions Caquetá-Putumayo, Box 25, n.f.

11 Although since the late 1990s migration associated with the *coca* economy decreased significantly, the population of the Putumayo continued to grow at a relatively fast pace due mostly to the recent oil and mining boom. The last national census (2005) estimates the total population of the department at 310,132, and the projection for 2015 is 345,204.

12 Agudelo to Minister of Public Works, 11th December 1932, AGN, MOP, Vol.3273, fols.199–214.

13 See, for instance, AGN, MOP, Vol.066, fols.612–621; Vol.1065, fols.250–251; Vol.1160, fols.137, 153, 162–169, 454, 578v; Vol. 4946, fols.495–498.

14 'Informe sobre la carretera Pasto-Puerto Asís', 3rd May 1942, AGN, MOP, Vol.1160, fols.162–168.

15 Leal to Baez, 3rd June 1942, AGN, MOP, Vol.1160, fol.169.

16 Lleras Pizarro to López, No. 990, 6th May 1939 (Franco Romo, Personal Archive)

17 Pseudonym requested.

18 'Informe sobre la carretera Pasto-Puerto Asís', 3rd May 1942, AGN, MOP, Vol.1160, fols.162–168.

19 'Datos sobre deslizamientos, derrumbes y erosiones en las carreteras', 17th November 1969, AGN, MOP, Box 6, folder 35, n.f.

20 'Estadísticas de accidentalidad carretera Pasto-Puerto Asís', Oficina seccional Invías Mocoa, Putumayo. Note: The Invías' data and quotes on road accidents presented in this section come from a paper file that is not paginated.

21 The Pasto-Mocoa road is 148 km long, but the section known as the *trampoline of death* extends from San Francisco (Km 73) to Pepino (Km 138).

22 For an account of the political evolution of the M-19 movement in the Putumayo see Ramírez *et al.* (2010, pp.62–64).

5

On the illegibility effects of state practices

In December 2009, almost 80 years after an engineer depicted the Capuchin road in the San Francisco-Mocoa section as an 'eyesore' and urged the government to totally replace it,[1] the following description of its substitute, the *trampoline of death*, appeared:

> There are serious traffic restrictions on the section of road (78 km) between San Francisco and Mocoa (capital of the Putumayo) built in the 1930s, which has long 4-metre wide stretches where only one vehicle can pass, high gradients, unstable areas, constant cloudiness, and sharp cliffs, making this one of the roads with the highest accident rates in the country (IADB 2009a).

This brief and technical diagnosis, one amongst many which for decades have accompanied this infamous road, promised however to put a definite end to it. Its source, a loan proposal from the Inter-American Development Bank (IADB) approved in that same month, contemplates the construction of an alternate road following roughly the same route the Capuchin missionaries chose a century earlier. The new road, originally expected to be concluded over a period of 8 years, will significantly shorten the distance (from 78 to 45.6 kilometres) and travel time (3 to 1.5 hours) between Mocoa and San Francisco. As for the *trampoline of death*, the project plans to turn it into an 'eco-touristic corridor'.

The IADB proposal makes no mention of the long history of struggles or the countless pleas and strikes demanding the replacement of the current

Frontier Road: Power, History, and the Everyday State in the Colombian Amazon,
First Edition. Simón Uribe.
© 2017 John Wiley & Sons Ltd. Published 2017 by John Wiley & Sons Ltd.

road. It does, however, make great emphasis on terms such as 'efficiency', 'safety', 'competitiveness', 'sustainable development', 'governance' and 'strengthening of the state', referring both to the project's chief objectives and the Bank's overall strategy for the country. Considerable weight is also put on the issue of economic and physical integration, both at the national and regional levels. The latter is explicitly linked in the document to the IIRSA (2011, Chapter 1), within which the road is considered to be a priority project, as it will allow interoceanic communication between the Atlantic and Pacific oceans through Brazil and Colombia.

Although the association of the project with the IIRSA has raised strong criticism from non-governmental organizations and citizen and indigenous movements about its extractivist logic and its potential social and environmental impacts,[2] for most Putumayenses the road represents a subject so sensitive that it often appears beyond dispute. A clear example of this is the tensions between indigenous and non-indigenous communities, where the former have been constantly accused of being 'enemies' or 'obstacles' to 'development', mostly due to their protest against the government's refusal to carry out a process of *consulta previa*.[3]

The expectations about the future road are many and diverse. At the more abstract level, the road is largely imagined as an infrastructure that promises to deliver 'progress', 'development' or even 'modernity', regardless of the varied and conflicting meanings and visions that these concepts tend to embody. At a more concrete level, it is conceived in terms of the tangible benefits it promises: shorter travel times and lower costs, greater comfort, and above all safety. For many, moreover, the road is synonymous with opportunity. Some have plans of starting up a small business on the side of the road or buying a piece of land for speculative purposes, even though the project includes several measures to avoid these activities. Others hope to get a job, although most of the available positions for locals contemplated in the project are the lowest paid and fall under the category of 'unqualified'. Politicians, meanwhile, regularly proclaim their great commitment to the project or even campaign by portraying themselves as its biggest supporters. Expectations, on the other hand, are not confined to the local level, and the most illustrative example is perhaps provided by a number of mining concessions granted to multinational companies within the project's area.

Throughout 2010, the year I lived in the Putumayo, those plans and expectations remained largely as mere speculative prospects or hopes, just as the road itself.[4] Nevertheless, if taken together, such plans and expectations comprise a multilayered cartography through which a single space was appropriated in multiple and often incompatible ways. This space, the relatively small area to be crossed by the road and its immediate surroundings, on the other hand, was not empty. Its inhabitants, largely *colonos* from Nariño, were estimated by one of the project's several evaluations in around 1,500 and classified as the population under the road's 'direct area of influence' (INCOPLAN 2008, vol.3). For them, although not devoid of expectations, the road represented primarily a source of everyday uncertainties about their present and future life. To a large extent, those uncertainties stem from the project's several safeguard policies, themselves to be translated into a myriad of land use and environmental measures and restrictions in the area. Yet what these uncertainties exposed was a much broader issue, namely, the project's rationale of rendering a target territory and population *legible* in order to make it *governable*.

This chapter addresses this subject of legibility in relation to the road project and has two main objectives. The first is to discuss legibility within the context of the project's goals and objectives. Specifically, I show how the project involved a series of conceptual premises and methods that despite their claims to objectivity and scientific validity did not so much produce accurate *contents* as privileged certain *forms* of knowledge and expertise while excluding or subordinating others. In this way, legibility is revealed not just as a *condition* required for the successful realization of the project's different goals but as a *means* to turn a specific territory and population into an object of governmental intervention.

The second aim is to consider legibility in relation to the gap between the project's 'theory' and 'practice'. I will focus on the project's attempts to clarify the land tenure situation in its area of influence, as it constitutes a clear example where this gap became particularly evident. Specifically, I illustrate how these attempts not only brought to the surface the state's deeply rooted or 'structural' legibility problems in the project's area but paradoxically ended up aggravating them. However, I will argue that this situation cannot be grasped only in relation to the issues the project identified but to those not taken into account or ignored, particularly people's customary land tenure practices.

As I will describe in the last part of the chapter, a significant aspect of those practices is that, as they accentuated the distance and conflicts between 'project' and 'community', they projected an image of these spheres as two isolated and diametrically opposite domains. However, a closer look at such practices will show that rather than drawing a neat line demarcating these domains, they evidence or reflect the multiple encounters and interactions between them.

'The illusion of transparency'

Roads have often been described as technologies of state power (see, for instance: Dalakoglou 2012; Fairhead 1992; Masquelier 2002; Selwyn 2001; Wilson 2004). This view largely stems from the fact that roads, regardless of the ends to which they have been conceived, constitute infrastructures that render territories and populations more 'legible' or visible to state surveillance and control. That this 'legibility effect' is in most cases far from absolute and in practice roads comprise spaces subjected to multiple subversions, appropriations and ends different to those envisaged, constitutes another subject. Ethnographical scholarship has provided compelling evidence on this particular issue, shedding light on the unstable, ambiguous and contested nature of such infrastructures (Campbell 2012; Harvey 2001; Harvey and Knox 2008; Kernaghan 2012).

The principle of legibility, on the other hand, is by no means confined to physical infrastructure and it has usually been conceived as an intrinsic aspect of state-making practices. A good example of an approach to the state from this perspective is Scott's (1998) well-known critique of twentieth-century socio-spatial engineering projects. Those 'great utopian' projects, which range from the Soviet collectivization schemes to the construction of Brasilia or the centralization of road networks in nineteenth-century France, are according to Scott bounded by a common history: they all resulted from a High-modernist ideology, whose ultimate objective was to make societies legible, no matter the cost, in order to render them governable. The principle of legibility, which in Scott's analysis is equated to Foucault's panopticon, appears as a radical, and in most cases disastrous, simplification (biological, social and spatial) of the world, whose underlying motives are 'appropriation, control, and manipulation' (Scott 1998, p.77). Furthermore, legibility, as noted by Trouillot, is not solely bound to the state apparatus and it can be better recognized as an 'effect' of both government and non-government practices and institutions (NGOs and multilateral agencies, among others), which implies 'the production of both a language and a knowledge for governance and of theoretical and empirical tools that classify and regulate collectivities' (Trouillot 2001, p.126).

This 'legibility effect' was implicitly present in the San Francisco-Mocoa road project, most patently in the assumption that the future road will foster 'greater State presence' and strengthen 'local governance mechanisms' (IADB 2009a, p.3). For the most part, however, and due to the project's extensive safeguard measures and its aim of protecting the environment on its area of influence, legibility was implied not as an anticipated *effect* but as a central *precondition* for its successful implementation. Thus, three exhaustive assessments covering the multiple technical, social, environmental and economic dimensions and strategies of the project were developed by different consulting firms prior to its final approval: the EAR (Strategic Regional Assessment), examining the project's potential environmental and social risks and opportunities at the national, regional and local levels; the EIA (Environmental Impact Study), which designed an environmental management plan aimed at minimizing the road's impacts upon its direct area of influence during both its construction and operation phases; and the PBMAS (Basic Environmental and Social Plan for the Protected Forest Reserve of the upper Mocoa river basin), aimed at defining the strategies and measures for the conservation of a Forest Reserve to be traversed by the road. These studies were subsequently synthesized in the PMASIS (Integrated Sustainable Environmental and Social Plan), conceived as the project's main management instrument to be followed and implemented by the different actors and institutions involved in the project.

Together, these assessments devised several measures that range from high environmental standards for the road design and awareness campaigns with local communities, to numerous control and supervision activities including surveillance by radar and satellite imagery, permanent rangers and checkpoints along the route, and telemetric monitoring of wildlife. What is of interest here is not so much the specificity and scope of such measures but the premise on which they are sustained. That is, that in order to be successfully implemented, they require and presuppose as a *sine qua non* condition a highly detailed knowledge of the area affected by the project. This condition might look obvious and elemental if seen from the planner's aims and perspectives. Not so obvious, however, are the underlying logics and paradoxes involved in the planner's discourse and practice. In order to illustrate this point better, I will focus on one of those assessments (the PBMAS), as it fully exposes not only these logics and paradoxes, but the way in which they relate to the subject of legibility.

As noted, the PBMAS study was conceived as a social and environmental plan for the zone to be directly affected by the project. It covered an approximate area of 70,000 hectares, even though it focused primarily on the Forest Reserve (comprising half of this area) to be dissected by the road, for which it devised four wide-ranging strategies: environmental land use; conservation and sustainable development; involvement of local communities; and operation, surveillance and control. These strategies derived from an extensive and thick assessment consisting of various volumes covering its different components (environmental, juridical, institutional, cultural, biophysical). The overall study was described as 'an unprecedented detailed scientific knowledge of the Reserve and its area of influence' (IADB 2009b) and, at least, if considered from the amount of data provided, this definition holds true. Through its numerous maps, charts, synoptic tables and texts, the reader can get a fair sense, sometimes excessive, of the Reserve's many aspects and variables, from precipitation levels to geologic and topographic analyses, geo-referenced native species, and its inhabitants' cultural traditions, average incomes and mortality rates.

'Detailed scientific knowledge', despite the objectivity this claim suggests, must not however be confused with accuracy, at least if we interpret this term as the faithful correspondence between facts on the ground and facts on paper. If we stick to this definition, the PBMAS study could be better portrayed, quoting Lefebvre, as conjuring up an 'illusion of transparency' or an illusion through which 'space appears as luminous, as intelligible, as giving action free rein' (Lefebvre 1991, p.27). Here, space is rendered intelligible not by means of fidelity but by concealing or ignoring those elements that are antagonistic or threaten 'transparency'. In other words, 'transparency' results not from the correspondence between reality and representation, but as the effect or the illusion that representation creates of reality, an illusion that, as Lefebvre argues, cannot be detached from the connection between ideology and knowledge.

Lefebvre's 'illusion of transparency', which he associates with the capitalist production of space, can be connected to Mitchell's remark on the nature of modern forms of calculation. The development of such forms, he notes, 'did not produce a more accurate knowledge of the world, despite its claims, nor even any overall increase in the quantity of knowledge. Its achievement was to redistribute forms of knowledge, increasing it in some places and decreasing it in others' (Mitchell 2002a, p.92). The most pervasive effect of this modern 'regime of calculation', together with the simplification, reformatting and translation practices it entails, as the author stresses elsewhere, is not that representation is detached from social reality but that 'social reality is to be ordered according to the principle of representation' (Mitchell 2002b, p.25).

The PBMAS's claims to 'scientific knowledge' can be better appreciated in this light, and the most striking example is perhaps provided by its 'Socioeconomic and Cultural Diagnosis' (INCOPLAN 2008, vol.3). Carried out mostly among the settlements in the road's area, classified by the project as the 'direct social actors', this diagnosis was defined as highly participative and inclusive, and its methodology included several workshops, household surveys and socialization meetings. Yet, and somewhat paradoxically, the same study listed at the beginning a number of pre-defined 'basic concepts' guiding the analysis such as 'community', 'participation', 'territory', 'social process' and 'sustainability'. Others, such as 'development' or 'sustainable development', are claimed to be 'collectively constructed' throughout the process, though they differ little from their standard definitions. This conceptual 'bias' becomes even more manifest in the section on 'peasant culture', where stereotypes of the peasant and the *colono* are widely reproduced. Thus, the peasant is persistently portrayed as bearer of a 'traditional culture', expressed in terms such as prevalence of self-subsistence economies, survival of folk beliefs or deep attachment to the land. At the same time, nevertheless, he is associated to a '*colono* culture', characterized mostly in negative terms. For instance, their arrival in the Putumayo at the times of the Capuchin mission is described variously as causing 'a break of the ecological balance kept during centuries by the pre-Columbian man', an 'alteration of the traditional production systems', and an 'uncontrolled destruction of forests' (INCOPLAN 2008, vol.3, p.20).

It is not the aim here to discuss the pervasiveness of stereotypes or tropes regarding the peasant or *colono* in official and academic discourse, or in the specific case of the PBMAS study.[5] Rather, the point I would like to stress is how, together, these tropes comprise a predetermined discursive framework or language through which the different project's 'social actors' are studied, surveyed and ultimately assimilated to it. Seen this way, the project's 'diagnosis' goal can be conceived not just as a way to render legible a specific territory and population, but to turn that same territory and population into an object of government. Not surprisingly, a continuous thread throughout the 'Socio-economic and Cultural Diagnostic' is the propensity to catalogue the Reserve's inhabitants simultaneously as a 'threat' and 'strategic allies' for the project. Hence, its emphasis (summarized under its proposed 'management model') on the relevance of 'governance mechanisms', from strengthening of local institutions to self-government education processes, and to the involvement of the local community in environmental management and surveillance measures for the Reserve (INCOPLAN 2008, vol.3, pp.76–80).

A first paradox resulting from this assessment exercise is that even despite the project's self-proclaimed participatory approach, its different outputs, *legible* to the planner, ended up being *illegible* to the project's 'stakeholders'. This situation manifested itself during the project's many 'socialization workshops' with the inhabitants of the Reserve.[6] The workshops that I attended during the time I lived in Mocoa unfolded in a similar way. They were all introduced by a group of professional technicians (agronomists, lawyers, sociologists and other professionals from related disciplines hired by Corpoamazonia, the project's main environmental authority) who presented a summary of the studies' results, in this case the

PBMAS, and the devised measures. Great emphasis was placed on the future land use and environmental restrictions within the Reserve, as well as the development of 'sustainable projects' with those to be affected by them. This point, understandably, was the one that always generated most tension, expectations and concerns among the participants. Usually, the professionals stressed, often referring back to the project's participatory approach, the importance of the community's initiative on the formulation of those projects. The participants, on the other hand, repeatedly manifested their lack of understanding or skills required to formulate such projects according to the established guidelines and procedures. The professionals, knowing by experience that this constituted a familiar issue with development projects, often found it difficult to reply. On some occasions, however, they came up with ingenious answers such as this one I heard from a specialist in project management who insisted the solution was to be found within the same communities:

> Don't be fooled. When the project's people come don't let them diagnose you: you have to say 'we are already diagnosed from within' ... You might say 'Ah, it happens that we are peasants and don't know about projects'. I tell you again: in many rural areas I've found with surprise illustrious professionals that know plenty. Why are they not involved? This is a question I have in mind (Socialization workshop, Mocoa, 11th June 2010).

The specialist's answer, though offering a 'solution' to the participants' concern, not only did not question the project's 'participatory' rationale but expressly reaffirmed it. This same rationale was differently expressed in the intervention of a peasant leader during the same meeting. Contradicting the same specialist's observation on the restriction of agricultural practices on lands exceeding a slope of 30%, he said:

> This is the point that must be discussed: that you may say one thing theoretically but the community knows another thing. Because I can tell: if the terrain is like this (steep) nothing could be done. But if the terrain is like this (flatter), there I can. So don't come and tell me that there I can't, because I know the slopes in this area are less steep. One thing is the norm and another what the people live and do. This has been the main problem: that the institution says one thing and sticks to it and we from experience and practice tell you another thing.

The crucial point here, as we can infer from the leader's objection, is not his perception of the planner's misconceptions on land use but how, under the project's rationale of 'participation', the former's experience rooted in 'practice' often ended up subordinated to the latter's expertise backed by the 'norm'. Yet, and despite the immanent logic embedded in this rationale – that reality is to be superseded by its own representation – the same objection also confirmed the breach between the realm of the norm and the level of practice or 'what people live and do'. This breach, which in the project would become especially manifest in its attempts to clarify the complex land situation within the Forest Reserve, would expose another paradox. For those attempts, as we shall see next, not only laid bare the state's severe legibility problems but ended up exacerbating them.

The Forest Reserve's cadastral 'confusion'

As mentioned previously, one of the project's core strategies was environmental land use. As part of this strategy various policies were devised, including environmental education, the adjustment of municipality land use plans of both Mocoa and San Francisco, and contingency plans to contain urban expansion and settlement along the road. Although these policies covered a wide area beyond the project's direct area of intervention, the main emphasis was on the Forest Reserve to be crossed by the road. In theory, this legal figure perfectly fitted the project's conservation goals, since in legal terms Forest Reserves constitute zones aimed exclusively at the rational establishment, preservation or use of forest areas. The Forest Reserve of the upper Mocoa river basin, on the other hand, falls under the category of 'protected', meaning that its only allowed activity is the extraction of 'secondary fruits of the forest' (non-timber products such as seeds, resins and fruits). Nevertheless, the law establishes that Forest Reserves can be subjected to partial subtractions in case of 'public utility' or 'social interest' activities. Such is the case of the road, 70% (30 kilometres) of which will be built within the Reserve. Thus, and in order to mitigate the road's impact in the area, the project not only designed several control and surveillance measures but contemplated an extensive management plan – the PBMAS –- to guarantee its effective conservation. The latter included, besides the aforementioned policies, the extension of the actual Reserve, from 34,600 to 65,300 hectares, under the consideration that this would consolidate a large conservation corridor by connecting different protected areas across the region.

In practice, however, things turned out to be much less straightforward. As noted, the implementation of all the proposed plans and measures for the Reserve required first an exhaustive survey, to be carried out by the PBMAS's different diagnosis components. These components, including the 'Socio-economic and cultural' described above, were mostly extensive descriptions of the Reserve's physical and human elements. The 'Legal Diagnosis' (INCOPLAN 2008, vol.4), aimed at examining both the different legal aspects related to the Reserve and its cadastral situation, was not different in this sense. Yet, if seen from the angle of its final outcomes, this diagnosis study did not bring any 'transparency', not even an illusion of it. On the contrary, what it revealed was the deep-rooted and structural illegibility of the state's own practices in the area.

The study in question identified a number of issues mirroring or accounting for this illegibility condition: the available official maps of the Reserve were of an inadequate scale to allow decision making regarding its management; the institutional presence in the area was scarce, hence translating into deficient control and surveillance; and its boundaries, having never being demarcated, remained largely unknown. Although serious, these issues were to be addressed during the project implementation stage.

A related and yet much more complex and difficult problem was the land occupation of the Reserve, a situation resulting from its many legal holes and inconsistencies. The first of these legal issues lay in the origins of the Reserve. Created in 1984, it had been conceived to preserve the upper basin of the Mocoa River with the aim of developing there a small hydro-electric plant for Mocoa and other Putumayo towns. However, this project never materialized, which according to the law means that the administrative act giving rise to the Reserve would have lost 'binding force' or practical effect after five years of its creation. The study, although acknowledging this situation, argued, by resorting to the country's intricate legal system, that it could be objected to by claiming that the expiration of such legal act was never officially declared.

But the legal conundrum of the Reserve did not stop there. A more difficult problem to settle was the fact that the Reserve was never registered in the 'Oficina de Instrumentos Públicos' [Public Instruments Office], a requirement that basically consists of a legal procedure to make public to third parties any 'limitation of domain' (e.g. land restrictions and forfeitures) over private or public property. In the case of the Reserve, the registration was essential to publicize the different restrictions pertaining to this specific protective figure, the most relevant of which constitutes the prohibition of land adjudication and settlement within its borders and surrounding areas. This issue was extensively addressed both in the PBMAS 'Legal Diagnosis' and in a latter separate assessment (IADB 2009c). Both reports, although analysing every possible legal way to solve this problem, could not conceal a critical paradox; that the Forest Reserve –unmapped, undemarcated and unregistered – not only remained a space illegible to the state, but the state, failing to observe its own legal codes, had itself contributed to exacerbate this condition. The latter report's observation on the complex land settlement of the Reserve plainly evidenced this situation:

> From the information gathered so far, there exists [within the Reserve] an amalgam of owners with titles, recipients of *baldíos* granted after the creation of the Reserve, possessors, some of them registered, amongst others (IADB 2009c, p.53).

This remark sought to summarize the conclusions of a cadastral survey carried out as part of the PBMAS 'Legal Diagnosis', which the same report judged incomplete and thus urged a new and more extensive one. Despite its incompleteness and 'inaccuracy' (explained partly as a consequence of the Reserve's indeterminate boundaries), a brief review of this survey sheds light on the state-assisted chaotic land occupation of the Reserve.

The survey examined 150 deeds, distributed in the ten *veredas* (rural settlements) overlapping the Reserve. Four main legal types of land tenure were identified: 'Plena propiedad' (absolute ownership) or lands originally granted by the government with public deeds and a clean chain of title (83 cases); 'falsa tradición' (false tradition) or lands whose titles are considered flawed or clouded, a situation generally resulting when someone who lacks 'absolute ownership' sells or transfers his land to another person (10 cases); lands with public deeds but a broken chain of title, a situation caused by different factors, including old deeds not listed in the Registry Office databases (12 cases); and occupied lands with cadastral numbers or deeds certifying not the land property but the 'mejoras' (land improvements such as crops and buildings) made by the tenant on lands claimed as *baldíos*, although in practice these are non-alienable as they are believed to fall within the Reserve (45 cases).

In theory, and due to the Reserve's category of 'Protected', private property of any sort should not exist within its boundaries. Thus, at the moment of its creation all existing private lands, possessions and 'mejoras' within the Reserve should have been expropriated or purchased by the government or, alternatively, land-use restrictions enforced, procedures that were never followed. Most alarming, however, was the adjudication of lands or the official recognition of any kind of claims to property following its creation, of which the study identified 28 cases among the aforesaid categories (excluding that of lands with a broken chain of title, for which the original date of adjudication could not be established).

A related and equally serious issue constituted the issuance of mining concessions overlapping the Reserve, some of them granted during the period in which the PBMAS was being established.[7] Although these concessions were at the time in their first stage ('exploration'), their mere existence, taking into account that mining is considered 'not compatible' with Forest Reserves' conservation status, constitutes just another example of the gap between norm and practice.

Both the mining titles and land deeds analysed by the 'Legal Diagnosis', though mirroring some of the material effects of the state's incongruous legal procedures, were to some extent readable or at least 'visible' to the project. Through their cadastral or land registration numbers they could be identified in terms such as location, area, proprietor/concessionaire and chain of title. Quite different was the case of the so-called 'private documents' or documents aimed at demonstrating possession or occupation of lands claimed as *baldíos*. These can also be of different types, yet the most common ones are 'contracts of sale' (written contracts formalizing land sales between seller and buyer) and 'extra-judicial statements' (out-of-court statements before a public notary declaring possession or occupancy of lands claimed as *baldíos*). Legally speaking, these documents are considered 'private' since they do not comply with the requirements or follow the basic legal or administrative protocols of 'public' documents: they are not inscribed in the Registry Office, do not have cadastral plans and land registration numbers, and in some cases are not even notarized. This does not mean, however, that they are 'illegal' or not recognized by public authorities. In many cases, for instance, they form part of the ordinary procedures for legal transactions of land such as sales or cessions. In others (e.g. land adjudication processes) they may serve as proof of tenure or possession of land. Still, these documents *per se* (e.g. without a land deed) do not have the same legal value of title deeds and, in the best case, may demonstrate possession but not ownership of land.

'Private documents', despite their extra- or semi-legal character, constitute a very common way to trade lands and are often used for different purposes such as evidence of tenure or access to government subsidies. Besides, sometimes their use can be associated to practices such as evading costs (sale or property taxes) or legal controls (e.g. lands that for different reasons cannot be formally traded). Moreover, and most significantly, these documents usually originate from real practices that do not necessarily take place outside or infringe the law (most of them are formalized before a public notary), and yet remain largely illegible to government functionaries. The 'Legal Diagnosis', for instance, identified and examined 259 such documents (against the 150 deeds) provided by people themselves upon request. The examination in this case consisted basically of a summary of the type of document and information regularly supplied by each of them (e.g. name of seller/buyer/grantor, area/name of land plot). This information is not homogeneous across documents: some do not register areas or names of land plots, others lack notary seals, while others are handwritten papers whose content is literally illegible. However, this aspect is irrelevant if we consider that no matter how detailed or partial, fictitious or real, current or outdated, this information – unregistered, unrecorded, and un-georeferenced – is, for its most part, indecipherable to the cadastral surveyor.

* * *

By the time the PBMAS study was completed, in July 2008, it was clear that the complex and in many cases indecipherable character of land occupation in the road's area of influence posed a serious challenge to the project's declared goals. Various alternatives to overcome this situation were proposed: partial subtractions of the Reserve's lands legally granted or demonstrating legal tenure; re-categorization of part of the current Reserve to a class ('Protected-Protective Forest Reserve') with lesser land-use restrictions; and purchase of lands legally owned and removal of tenants 'illegally' occupying the Reserve (e.g. with no titles or titles deemed invalid). The problem, still, was that regardless of the alternative or combination of alternatives to be applied, in order to be viable, any option required first a clear understanding of the Reserve's cadastral situation on the ground. The 'Legal Diagnosis', as shown, rather than solving this problem, plainly exposed the Reserve's illegibility. It was on these grounds that a new and more exhaustive survey was recommended, aimed at updating and complementing the existing data in both the current Reserve and its projected expansion area.

Two years later, in 2010, this new survey had not been concluded and the Reserve land problems, far from being resolved actually seemed to have worsened. This situation, on the other hand, and somewhat paradoxically, was partly fostered by the project itself. With the expected starting date of the road works approaching, the project officers urged people to provide their land deeds or documents, while repeatedly stressing the prohibition of land transactions while the Reserve's legal situation was not clarified. It was evident, however, that the officers themselves were incapable of providing definite answers to many of the people's uncertainties and concerns about their future: Were their lands going to be expropriated by the government? Were they going to receive any compensation? Exactly which lands fell within and outside the Reserve? What would happen to those with invalid titles?

Unanswered or given ambiguous responses, these uncertainties ended up causing or increasing distrust among the inhabitants of the Reserve. Various project officers I interviewed acknowledged this problem, and observed that in some *veredas* people had refused to provide their land titles. The version I got from a woman peasant leader was no different, although it made clear how the project's many studies were responsible for the current tensions between project workers and local communities. 'No study has managed to make things clear. None', she said while complaining that during the past three years people from her *vereda* had handed in their land titles four times, but so far had not received any reply. And then she went on describing the routine practice: 'the lawyer comes and asks for copies of titles and then he gets lost and never shows up again. And then comes another study and a different lawyer and so forth. Because they continue to have that confusion'.

'That confusion', promoted by the state's incongruous legal practices and further accentuated by the project itself, exacerbated people's uncertainties and distrust, which in turn aggravated the Reserve land problems. This sort of vicious circle had become evident during 2010, when rumours about people selling or buying lands within the Reserve began to circulate with some regularity. Although at the time most of these rumours seemed to be just rumours, some did actually materialize. For example, a logger from one of the Reserve *veredas* told me he had recently sold, upon hearing that timber extraction would be strictly forbidden, half of a 100-hectare farm he owned located 'right in the centre' of an area he'd heard would be crossed by the road. The sale, ironically, was facilitated by the fact that his farm did not have deeds but an old 'contract of sale' through which he had originally acquired the land. This transaction, as others I heard of at the time, done through private document, would inevitably add more confusion to the already chaotic cadastral situation of the Reserve.

By the end of 2011, the proliferation of land sale/buy signs posted along the project corridor (Figure 5.1) showed how this situation had

Figure 5.1   Requests to purchase and sales of land plots in the road project site (December 2010).

intensified. These signs, though highly visible to the project, stood for deals that in most cases were illegible to it and hence beyond its control.

The crucial aspect of this 'illegibility effect', I suggest, is not the ways in which it exposed the gap between 'norm' and 'practice' or 'project' and 'community', but rather how this gap reflected the multiple entanglements between these apparently detached or independent spheres. In other words, as it was emphasized, the 'peasant's indecipherable' land settlement practices could only be explained in connection to the government's normative inconsistencies, just as people's uncertainties and distrust could not be isolated from the project's cadastral 'confusion' and the 'diagnosis' methods through which it attempted to 'clarify' it. In this order of things, it is no longer surprising that the project's surveys did not only fail to overcome the land tenure issues they were confronted with but actually aggravated them. However, as I shall describe next, to fully grasp this illegibility effect, to understand its roots and how it is reproduced in time and space, we have to look not just at what the project 'saw' but also at those elements that from the beginning it ignored or, confined to its conceptual framework, was unable to perceive.

Becoming illegible: a short case study

Campucana is one of the ten *veredas* that, according to the project, totally or partially overlapped with the Forest Reserve. It is located about seven kilometres north-west from Mocoa following the old Capuchin road and it is one of the *veredas* to be dissected by the future road. The PBMAS estimated its population as 165 individuals distributed in 36 families and, based on the study's surveys and workshops, summarized the main problems of the *vereda* as the loss of cultural traditions and values, lack of access to local and regional markets, and lack of technical support in agriculture, the *vereda*'s main economic activity. While these problems differed little from those identified in the other *veredas*, the land tenure situation was among the most complex ones, as the study found that only 20% of the surveyed land plots had title deeds, while the missing 80% had only private documents, mostly 'contracts of sale' (INCOPLAN 2008, vol.3, pp.48–42, 78).

The PBMAS 'Legal Diagnosis', containing separate annexes for each *vereda* aimed at examining land titles and documents individually, found in Campucana most of the land issues commonly identified in the project's area: various land plots that the field surveys situated within the *vereda* but the titles located in others; lands granted after the creation of the Forest Reserve; and several cases of partial sales of plots with public deeds through private documents, some of which did not register areas or whose content was illegible (INCOPLAN 2008, vol.4). If the detection of these issues exposed the intricate land tenure patterns within the *vereda*, the PBMAS did little to explain them. For instance, the brief chapter on Campucana from the 'Socio-economic and Cultural Diagnosis' addressed the subject of land use in the *vereda* by simply observing that 'the land tenure is legalised through public deeds and others [sic] by contracts of sale' (INCOPLAN 2008, vol.3, p.49).

In order to understand the origins and development of such patterns and its illegible character, we should look back to the colonization history of this particular *vereda*, a story that goes back to the early times of the Capuchin road. As previously noted (Chapters 2 and 3), the missionaries tirelessly encouraged the migration of *colonos* to the region by highlighting the abundance of *baldíos*, a strategy that was also employed to attract workers for the road. Expectedly, the lands most coveted by those early settlers were those located in nearby populations or along the road. The section of the road from the Sibundoy Valley to Mocoa (the area comprised today by the Forest Reserve) seems to have been an exception, a situation associated with the rugged topography of the area. For instance, in June 1912, a few months after this section had been concluded, the Inspector sent from Bogotá to examine the works observed that in most of this section the road crossed 'totally virgin and deserted forests, through the cliffs and slopes of the cordillera'. This situation, however, changed noticeably from the actual site of Campucana to the town capital, where the Inspector found 'arable lands of barely fair quality, all claimed as *baldíos*, and all granted provisionally by the township of Mocoa'.[8]

Although Campucana was then a mere *paraje* (stopping place) on the road and the origins of the actual *vereda* date from more than two decades later, the Inspector's observation indicates the early land speculation in the area. However, government regulations regarding *baldíos* were at the time rather vague and unclear, and it is difficult to establish how many of those land grants survived or eventually translated into title deeds. The eldest inhabitants of the *vereda*, for their part, locate the arrival of the first settlers in the mid-1930s and 1940s, and the oldest land titles and documents found by the PBMAS date from this period.

As in the other surrounding *veredas*, Campucana's first settlers were mostly families from Nariño in the search of cheap or vacant lands, already scarce in the Sibundoy Valley. Yet, what made these lands attractive was not only the perspective of claiming them as *baldíos* but its abundance of precious woods, a highly valued product that for several decades became the main means of livelihood of the area. The proximity of such lands to the Capuchin road, still in use at the time and kept up by the settlers themselves, facilitated the access to timber, transported by mules to Mocoa and San Francisco.

The extraction of timber in Campucana, an activity that continues to play a relevant role in the economy of the *vereda* even though it constitutes a practice long deemed illegal, is vital in understanding its complex land tenure system. Part of this complexity has to do with the simple fact that land, especially in the early days of the *vereda* when *baldíos* were still abundant, was valued not so much in terms of size but the amount of timber it contained, and above all its access. This aspect is mirrored in the old land deeds and documents, which rarely register areas but instead emphasize the proximity of the lands to the road and in some cases mention the existence of timber. At the same time, most of these lands were not enclosed and its borders were frequently footpaths, creeks, trees, stones and other landmarks largely unintelligible to outsiders. The following description of a land plot coming from a private document offers a good example of this situation. The document, a notarized contract formalizing the sale of a land plot in the site of 'El Conejo' – a creek that runs through Campucana – in August 1955, does not register the plot's area or name but instead contains the following detailed description of its borders:

> To the East, it borders with the bridle path that goes to San Francisco, this side measures eighty (80) metres; to the North, in West direction, borders with the land of Hermógenes Nupán, a small creek upwards, until finding a guava tree and a plant of chontaduro [*Bactris gasipaes*, local palm], this side measures seventy two metres; from this landmark it goes South, bordering with the same Hermógenes Nupán, until finding a *huecadita* [small ditch], this line measuring fifteen metres (15); from this landmark in the same West flank, goes back to the East until it finds a *barbasco* tree,[9] this line [sic] twenty metres (20); from here, forming a set square, it goes South until its finds an *ojo de aguas vivas* [natural water source] this line measuring thirty (30) metres; from this landmark, in the West flank, heads again to the East, following the stream of the *ojo de agua* mentioned above, bordering with lands of Benedicto Nupán, until it gets again to the above quoted road to San Francisco, point of departure, shore of the waters above referred (INCOPLAN 2008, vol.4, cadastral annex vereda Campucana).

According to the seller's version recorded in the same document, this land plot was a *baldío* inherited from his father, and it is quite likely that it would be later granted by the government to the current or future buyer. In such a case, although ownership was difficult to determine as the document does not seem to be connected to any land deed included in the study, it would eventually become a property with title deeds, containing more 'legible' data such as name, area and cadastral plan. However, even considering this to be the case it is predictable that this 'legibility' would be rather ephemeral, as it is also very likely that the same land plot would be later subjected to partial sales and other trans-actions with private documents, a common practice in the *vereda*. This illegibility pattern only accounts for a small part of the tenure practices in the *vereda*. Beyond documents and deeds, there is a multitude of the so-called 'arreglos de palabra' (verbal agreements) through which lands are regularly appropriated, traded, inherited and rented. A typical example of these agreements constitutes the tenure of lands that people in the *vereda* call *baldíos*, though in practice are non-alienable lands for different reasons (e.g. they are within the limits of the Forest Reserve or its surrounding areas). These lands are usually the more remote or less accessible ones, for the most part uninhabited or only sporadically occu-pied. Still, and even though they lack any kind of document, they all have 'owners'.

In order to illustrate this point better, I will use, as an example, a farm from the *vereda*. Originally a proper *baldío*, this farm was granted in 1982 to José Pérez,[10] its current owner, after he had occupied it for more than ten years. It extends to 25 hectares, which is above the average size for the *vereda*, and the PBMAS 'Legal Diagnosis' classified it as a land plot in 'absolute ownership'. Differently to most farms, it maintained its original size and no total or partial sales are registered. For several years – during which José made a living from the extraction of timber – it remained unfenced, its borders with the neighbouring farms consisting of landmarks such as those mentioned above; nowadays, however, as he keeps a few cattle, it is partially fenced with barbed wire. The plot is divided as follows: two and a half hectares in coffee crops (the main cash crop) mixed with staple crops (manioc, plantain, fruit trees), five hect-ares in pasture lands and eight hectares of *rastrojo*.[11] The rest of it, about ten hectares, consists of the so-called 'monte' (secondary forests too steep to be farmed but considered an important source of different timber and non-timber products).

The above description provides a basic picture of the *baldío* granted by the government to José. In practice, nevertheless, José's farm extends far beyond the 25 hectares he legally owns or that are registered in the title deeds. The reason is basically that the farm's southern boundary borders with an extensive area of forests, whose main access is through the farm. It is from this forest, whose area José estimates at about 100 hectares, that he extracted cedar, cumin and other precious woods for more than three decades. However, as José himself acknowledges, the basis of 'ownership' in this case is, if not ambiguous, contingent at least. Thus, while he assures that the only 'legitimate owner' of this forest is the state, he also argues that as long as the state does not claim it, it will continue to 'belong' to him.

Even nowadays, when José no longer logs timber, his farm continues to be the main 'gateway' to the forest and anyone wishing to enter it needs his permission. Yet, it would be erroneous to think that his 'ownership' of this forest is based solely on the fact that he controls the access to it. Actually, what ultimately defines 'ownership' here, as in many other informal land tenure arrangements, is the fact that, despite lacking any written proof of property or possession, it is acknowledged and respected by the other inhabitants of the *vereda*. In other words, 'ownership' based on 'access' constitutes here one of the many different customary norms that together sustain and regulate the life of the *vereda*.

A central aspect of those norms and the infinite arrangements stemming from them is that, while they are perfectly legible in the eyes of the *vereda's* inhabitants, they remain largely invisible or indecipherable in the eyes of the state. The expression of frustration from one of the project's professional technicians dealing with the issue of clarification of land tenure within the Reserve is quite telling in this regard. In explaining to me the hardships and conflicts he and his team had been facing in the process, he said:

> How can I, as State, *sanear*[12] an area that you, as *Colono*, claim that you have been making use of for 20 or 10 years, but it does not have a title, nor a contract of sale, nor anything but you say you bought it from someone else for two hundred pesos twenty years ago? For the fact is that all the lands within the Reserve have owners. And one goes and looks and they say 'this belongs to that person because there is this *trocha* that goes in there and it was built by that same person', or 'this land right there belongs to him because he came here first and logged some timber and opened some *trochas* so this belongs to him'. These actions have created certain forms of understanding and certain land tenure agreements among people. But then what does this mean for Mr. State and Mr. Government when they go and try to clean this up? It means that there is going to be a conflict (Personal interview, Mocoa, 8th July 2010).

At first glance, this observation may suggest that 'state' or 'government' and 'community' stand as two separate and in many ways antagonistic orders, the latter governed by custom and the former by law. Yet, as it has been repeatedly shown, this is far from being the case. Certainly, many informal land tenure arrangements cannot be isolated from the practices of the state. Sometimes, they take place in response to the latter's coercive powers. This is the case, for example, of land transactions through private documents or the undervaluing of land in public deeds as a way to avoid land taxes, a widespread practice in the *vereda*. Still, at other times, those arrangements may also result from or reflect people's strategies to take advantage of government policies. An example of this, worth describing in some detail, is the 'Programa Familias Guardabosques' (PFGB [Forest Ranger Families Programme]), a government-run conditional cash transfer programme that is well remembered in Campucana.

PFGB was launched by President Alvaro Uribe in 2003, and was essentially aimed at eradicating illicit *coca* crops in the country, targeting rural populations that were directly involved in *coca* farming or, as in the case of Campucana, were considered to be at risk.[13] During the period 2002 to 2010 the PFGB reached a considerable part of the Putumayo, at the time one of the country's departments with the highest presence of *coca* crops (UNODC 2008, p.13).

PFGB arrived in Campucana on July 2006, during Phase III of the programme, and extended for three years. Although the influence of *coca* crops in the area was negligible if compared with other parts of the region (UNODC 2006, pp.75–77), the programme generated big expectations in the *vereda*, mostly due to the money incentive. This consisted of a bi-monthly subsidy of 600,000 pesos (approx. US$200), 65% in cash and 35% as a 'saving' to be withdrawn at the end of the programme and invested in a 'productive sustainable project' (popular ones in the *vereda* included coffee, cattle, poultry and sugar cane). As for the eligibility requirements, apart from the commitment to not growing *coca*, they consisted basically of having family, being head of a household and demonstrating ownership or possession of a farm of at least two hectares.

Not surprisingly, everyone in the *vereda* wanted to be a beneficiary and those who were not eligible resorted to all kinds of strategies to fulfil the programme requirements. Isabel, a community leader from the *vereda* who worked as programme overseer, recounted some of them: widows who 'hired' partners for a percentage of the subsidy, singles who arranged fictitious marital unions overnight to pass themselves as couples, and others who temporarily 'rented' children. As these 'special' cases were first internally evaluated by the *Junta de Acción Comunal* (the *vereda*'s community board, responsible for reporting to the programme office in Mocoa), conflicts regarding who should or shouldn't get the subsidy were common. Furthermore, those who were refused by the Junta often opted to act without its consent, and threats to the community overseers to cover them up were not infrequent.

Strategies to comply with the requirement of land ownership/possession proliferated as well. The deeds and documents gathered by the PBMAS 'Legal Diagnosis' (carried out at the time the programme was being implemented) represent an important source in this respect. During 2005 and 2006, and especially in the months preceding the arrival of the programme to the *vereda*, the increase of both legal and informal land transactions (extra-judicial statements declaring possession of *baldíos*, land donations and total or partial sales of lands with legal titles through private documents) is quite noticeable. The PBMAS makes no mention of the possible relation between those transactions and the PFGB, and since land documents were to be supplied by people, it is more than likely that there were many others impossible to trace. Isabel, for instance, mentioned that with the arrival of the programme 'the documents soared' in the *vereda*, and she recalled different cases: a person who owned a farm exceeding the two hectares required by the programme had 'lent' land parcels (through notarized contracts of sale) in exchange for a fraction of the subsidy; an extended family living under the same roof who split their farm in various pieces and built separate homes to appear as separate owners; a married couple who divorced and split both the farm and the children so each of them could present themselves as head of a household and receive two subsidies.

But Isabel's stories drew attention not only to the multiplication of informal arrangements to enter PFGB. More relevant was how many of these arrangements, despite their fictitious and provisional nature, ended up having real effects. Her own experience with the programme provides a telling example in this respect. Her father, in order for all his sons to be eligible, 'issued' private documents (through contracts of partial sale of his own farm) to those who did not own land. Still, he was emphatic and told everyone that such documents were only for them to obtain the PFGB benefits, for his will was to eventually divide his farm in equal parts among all his sons. Yet he died months before the PFGB came to an end and when this happened, by the end of 2009, a long dispute arose among brothers concerning the documents in question. According to Isabel, the main source of the conflict, still unsettled, was that the four brothers who had been given documents (this was not her case since she was married and owned a land plot prior to the arrival of PFGB) had refused to surrender them, alleging that the old man had given the land to them as a gift, and thus argued they were its 'legitimate owners'.

Like Isabel's, there are several other stories of land deals and agreements related to the PFGB that ended up altering existing land tenure arrangements. In some cases these alterations resulted from verbal agreements broken by conflicts. In others, they were just the consequence of provisional arrangements that became permanent, or even of real land transactions made with the subsidy money. A significant aspect of such arrangements is that they brought more complexity to the *veredas'* already intricate land tenure system, to the point they *projected* a dual image of 'community' and 'government' or 'state' as two totally detached and dichotomous orders. In the remaining part of the chapter, I emphasize this aspect by contrasting two cadastral maps of Campucana (Figures 5.2 and 5.3). In doing so, however, my primary aim is not to use

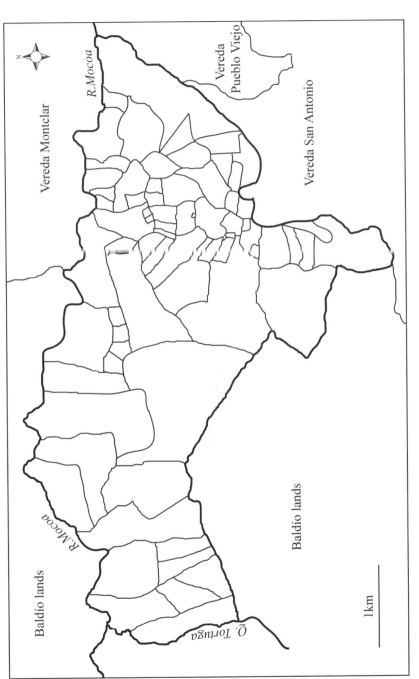

Figure 5.2 Campucana's official cadastral map, 2004.
Source: Elaborated by author. Cartographic base: IGAC National Cadastral System, map sheet ID number 430-II-C and 430-II-D.

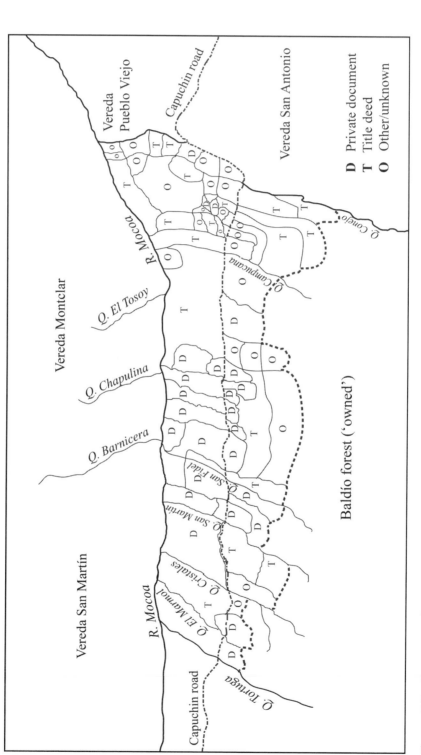

Figure 5.3 Campucana's peasant cadastral map, 2010.
Source: Digitally traced by author. Based on map drawn by Ernesto.

the maps as further evidence to highlight this dichotomy or gap, but rather to deconstruct them in order to draw attention again on the multiple imbrications and entanglements between the spatial and social orders they seek to capture.

Entangled maps

Figures 5.2 and 5.3, as indicated by their titles, correspond to official (government) and peasant cadastral maps of the *vereda* Campucana.[14] Apart from a very few perceptible similarities (a bigger concentration of land plots on the right-hand side and a partial correspondence in the location of neighbouring *veredas*), the two maps have little in common: the location, size and shape of land plots are different, the cadastral and geographical 'data' displayed on each of them is of a different sort, part of the *vereda*'s adjacent areas are identified differently, and the contours of the *vereda* are markedly dissimilar. Even though these discrepancies might largely reflect the aforementioned gap between the state's 'norm' and people's 'practice', they are not confined to this gap. Another important factor is the different ways in which peasant and state project or translate 'reality' from ground to paper.

The first map is a copy of the official or state cadastral plan of the area comprised by Campucana taken from the Instituto Geográfico Agustín Codazzi's (IGAC [National Institute of Geography]) National Cadastral System, the government system that gathers cadastral information at a national level. There are various reliability and accuracy issues that could be mentioned in relation to this map. In the first place, the cadastral information is outdated. Although in theory the national cadastre has to be updated every five years, the data displayed in this map is from 2004, and in 2011 the upgrading process had not been carried out in the Mocoa area. This means that all the formal land transactions (through title deeds) than had taken place since that year, including those examined in the PBMAS 'Legal Diagnosis', are not shown in the map.

More strikingly, this map still shows as 'baldío lands' part of the areas surrounding the *vereda*, areas which according to the PBMAS not only fall within the Forest Reserve created in 1984, but were actually granted by the Colombian Institute of Agrarian Reform (INCORA, replaced by the Colombian Institute of Rural Development (INCODER) in 2003) prior to and after this year. Moreover, the borders of the *vereda* displayed on the map only partially correspond with the actual or 'real' borders recognized by its inhabitants. This situation is not so marked in areas where boundaries are fixed by natural landmarks that change little over time as in those where limits are imaginary lines, generally defined by the borders of the land plots themselves. These boundaries are seldom static and change sporadically, mainly as the result of adjudication of lands beyond the *vereda*'s current boundary or land transactions of bordering lands (e.g. when someone from the *vereda* sells or buys a land parcel bordering with another *vereda*).

Although an updated version of Campucana's state cadastral map would overcome most of the aforesaid issues, this new map would still be far from resembling the peasant's cadastral map. The reason, as previously noted, has to do not just with the peasants' myriad informal and semi-legal practices carried out without the oversight of the official surveyor, but also a fundamentally different process of map making. In order to illustrate this point further, it is worth describing in some detail the peasant cadastral map.

This map (Figure 5.2) is a copy of a handmade map drawn by Ernesto, an inhabitant of Campucana. Ernesto was around 60 by the time I met him and was one of the people with a good deal of knowledge of the area. He had lived in the *vereda* for 42 years and, as most men of his generation, had made a living mostly off timber and other extractive and agricultural activities. However, for many years he had also worked sporadically as a *trochero* (trail opener) in the numerous road-engineering studies undertaken since the early 1980s. Thus, he knew the Forest Reserve inch by inch and could talk endlessly about its intricate tenure patterns, including the history of most of the *fincas* (farms). It was listening to his minute descriptions that I eventually proposed to hire him as cartographer, so he might put some of his *trochero* knowledge on paper. The task, we soon found out, proved far more complicated than initially envisaged.

As the main purpose of the exercise was to contrast the official cadastral plan of Campucana with Ernesto's map, I provided him with an enlarged copy of the IGAC's map and suggested that he drew on top of it using a distinctive colour. Although the basic idea was to overwrite the original plan in order to highlight similarities and differences, it proved totally fruitless. Not only did the cadastral plan seem totally alien to Ernesto, but its lack of basic geographical data made it impossible for him to orientate himself on the map. Consequently, we decided to try again with a physical map of the area showing rivers and contour lines. This time Ernesto could roughly locate the *vereda* following some of its natural boundaries shown in the map, primarily the Mocoa River. However, he again found the conventions unfamiliar, and despite turning the map upside down several times, he could simply not find his way on it.

After two failed attempts, it seemed clear it was impossible for Ernesto to 'find himself' on any official map. Yet we agreed to make a third and last attempt. A few days later, I came back with a couple of large sheets of blank paper for him to start a new map from scratch. It took him some time to think of how to start, but eventually he began by drawing two horizontal parallel irregular lines across the paper, which he identified as the Mocoa River and the Capuchin road. Subsequently, he traced a series of perpendicular lines representing the main tributary creeks of the Mocoa River. Once he had established these landmarks and natural boundaries, the task of mapping the layout of the *vereda* and the land parcels seemed far easier. It was easy to see that he had found his way on the map, and while he drew it was almost as if he had again assumed his role of *trochero* and had embarked himself on an imaginary journey through the *vereda*. Here and there he spotted an old tree or abandoned path, located a farm and introduced its tenants or tenure 'status' (e.g. private document, title, other), pointed at a *baldío* forest and named its 'owner', and so on. And the more pieces he put together, the more intricate the map became, to the point that at the end it resembled an entangled puzzle of names, landmarks and intertwining lines and arrows.

The figure displayed here is a simplified or 'readable' version of the original map, showing only a few conventions depicting some basic features of Campucana's land tenure system according to Ernesto. We could not say this map is more 'accurate' than the official cadastral plan, at least if we stick to the strict definition of the term. For instance, it does not have a scale, the boundary and landmark lines are arbitrary strokes rather than topographic projections, and the shape of land polygons are more graphic than realistic. Yet, if we measure cartographic accuracy not as the degree in which the features on the map conform to facts on the ground but the extent in which the map discloses the *nature* of such facts, we could then argue that Ernesto's map is far more accurate than the state map.

Just to provide an example of this sense of accuracy, let us consider the way in which the southern boundary of the *vereda* is represented on both maps. In the IGAC's plan, this boundary is signalled by a neat line defined by the legal limits of those land plots bordering with the area classified as '*baldío* lands'. Despite the expected margin of error in measurement and the fact that it is outdated, this boundary line could be said to be a fair projection of the *vereda*'s southern border, that is, so long as we assume that the legal limits of the land plots conform to their actual ones. Yet this is rarely the case, particularly in areas where land parcels border with *baldío* lands such as the farm described in the previous section. This gulf between actual and legal or official boundaries is, in contrast, clearly expressed in Ernesto's map. In this map, there is no such thing as a border line but just an irregular and dotted trace whose purpose is precisely to stress the undefined, porous and shifting nature of boundaries. Accuracy, in this case, is then not an effect of measurement but, conversely, the art of expressing the incommensurability of certain facts on the ground.

'The art of being governed'

Even if we translate or decode the state and Ernesto's maps in order to render them comparable, as I attempted here, the fact remains that they are highly incompatible. The greatest proof of this is perhaps that the IGAC's cadastral plan is as illegible to Ernesto as his own map would be indecipherable and useless to the cadastral surveyor. If we transfer this mutual indecipherability from the maps to the spatial orders they seek to represent, one would be highly tempted to conclude that these orders are equally incompatible: a 'state space' governed by panoptic transparency and control, and a 'non-state space' dominated by messy and intricate practices hampering, subverting and evading such transparency and control.

James Scott, who has extensively explored the subject of legibility in the context of state formation, has stressed the crucial importance of such spatial dualisms to understand dynamics of domination and resistance in the colonial and postcolonial world. In a book suggestively titled *The Art of Not being Governed* (2009), Scott has looked closely at this dualism in the specific context of upland Southeast Asia, focusing especially on the populations that have successfully managed to remain 'stateless' for two millennia. This book can be read in many ways as the counter story to Scott's (1998) *Seeing like a State*, introduced earlier in the chapter, for its main concern is not about the state's ruthless technologies of legibility but the multiple elements and practices (geography, agricultural techniques and strength of oral traditions, among others) defying domination and control and thus facilitating the existence and survival of 'non-state' spaces. More importantly, he insists that his study is not solely concerned with a particular region or people and is more generally conceived as a 'fragment of what might be properly considered a global history of populations trying to avoid, or having been extruded by, the state' (Scott 2009, p.328).

Scott's call for a global history of state avoidance, which he encourages by mentioning similar stories of other peripheral regions and populations world-wide, is relevant as it seeks to challenge traditional narratives of state-building processes. However, this view poses serious limitations to the study of spaces where dynamics of domination, subversion or resistance cannot be isolated but rather coexist and are deeply intertwined. This is the case of Colombia's internal frontiers, which as I have argued cannot be conceived as a 'stateless' territory but rather as spaces whose inclusion in the state has historically been dependent on a series of exclusionary practices. Accordingly, I emphasized that we cannot assume that 'state' and 'frontier' exist as detached and homogeneous realms, the former being synonymous with domination and the latter with evasion or resistance. This means that in order to understand how domination, resistance, subversion and, more broadly, hegemony works, it is essential to account for the multiple entanglements and interactions through which both state and frontier have been discursively and materially constructed.

The land conflicts associated to the San Francisco-Mocoa road project discussed in this chapter shed important light on these interactions and entanglements. As it was argued, the project's goal of legibility was related to the broader governmental rationale of fostering state presence and control in the region. This rationale cannot be isolated from the inclusive-exclusive relationship between frontier and state. A clear example of how this relationship is manifested in the context of the road project constitutes the project's claims to 'participation' and 'inclusion' of 'stakeholder' communities (communities that were already classified and coded under categories such as 'peasants', 'indigenous' and 'colonos'), which in practice revealed how certain forms of knowledge were privileged and others subordinated or marginalized.

Still, this legibility goal, at least in what concerns the issue of the land tenure situation on its area of direct influence, largely failed. Although this failure was not solely a consequence of the project's logic and rather exposed the state's structural legibility problems in the area, this logic contributed to exacerbate these problems. Illegibility, moreover, was not just a 'state' or 'project' effect, but also resulted from a wide array of everyday land tenure arrangements and tactics. As it was illustrated in the section on the *vereda* Campucana, these arrangements and tactics were largely part of people's customary norms regulating the tenure and use of land. In many cases, however, they took place as a way of avoiding taxes or with the purpose of appropriating non-alienable lands. Still, at other times, as in the case of the PFGB, they were part of people's manoeuvres to take advantage of government subsidies, or even resulted from family and neighbour conflicts occurring in the course of the programme.

But the crucial aspect of those arrangements is that they constitute spatial practices that do not exist outside the state, let alone mirror a 'non-state' space. In other words, whether they respond to surveillance or coercive policies or result from strategies or manoeuvres to take advantage of government schemes, the fact is that they are not 'external' to them. Rather, they resemble de Certeau's 'tactics', or those innumerable procedures that defy or subvert disciplinary power and yet are carried out by 'groups or individuals already caught in the nets of "discipline"' (De Certeau 1988, p.xiv). Put in a somewhat different way, these everyday practices, though informal or extra-legal in nature, not so much embody 'the art of not being governed' as 'the art of being governed'. This distinction is crucial as it undermines the rigid divide between domination and resistance and, in the specific context of the Colombian Amazon, helps understanding the frontier–state relationship in dialectical rather than dichotomous terms.

Notes

1 Rafael Agudelo to Minister of Public Works, 11th December 1932, AGN, MOP, Vol.3273, fols.199–214.
2 For a detailed account of criticisms to the project see Flórez (2007) and Salazar and McElhinny (2008).
3 The *consulta previa* (previous consultation) constitutes a fundamental right held by indigenous peoples and other ethnic groups that basically requires them to be consulted in the case of infrastructure projects that might affect their territories. For a detailed analysis of the conflict around the *consulta previa* in the context of the road project see Chaparro (2015).
4 The works, initially expected to start by May 2010, did not begin until February 2012. In October 2016, only 15 kilometres (out of 45) had been built.
5 For a detailed discussion on this subject in the specific context of the Putumayo and Amazon regions see Chaves (1998); see also Ramírez (2001, pp.44–52).
6 These workshops constituted a later stage of the project and, as their name suggests, were mostly aimed at disseminating its projected strategies and measures in the short and long term.
7 The project identified four mining titles (in exploration phase) issued during 2007 and 2008, encompassing a total area of 7,830 hectares, 23% of which overlapped the Reserve (5.39% of its current area and 7.21% of the projected expanded area). In addition, it found a large oil concession (in prospection stage) overlapping 21.3% of the current Reserve (IADB 2009b, pp.144–145).
8 Rufino Gutiérrez to Ministry of Public Works, 24th June 1912, AGN, MOP, Vol.1407, fols.579–582, fol. 581.
9 Generic name for plants containing poisonous chemical compounds commonly used for fishing.
10 The names used in this chapter are pseudonyms.
11 *Rastrojos* are secondary-growth forest or woods resulting from lands cultivated in former times, which are left fallow for 2–3 years.
12 *Sanear* is a term which in the context of land tenure means the clarification and cleaning up of irregular or clouded land titles.
13 For a general description of PFGB see UNODC and Acción Social (2007).
14 The inhabitants of the Forest Reserve often identify themselves as peasants and the use of the term here must be understood in this context rather than as a conceptual category applied by the author to this population.

6

The politics of the displaced

In the preceding chapters, I introduced a wide range of characters who in different ways relate to the never-ending project of road building in Colombia's Amazon frontier. Together, these characters could be conceived or read as many different fragments and voices, past and present, of the same story. My aim in this book, however, has not been to assemble those voices and fragments into a single narrative of a particular infrastructure, but to place them in the spatial and historical process of frontier-making in the Amazon. This process, I have argued, has been primarily shaped by a relationship where the assimilation or incorporation of the frontier to the spatial and political order of the state has historically depended on its exclusion from the imaginary order of the nation. In doing so, I have attempted to present the road's many characters, from early twentieth century statesmen and missionaries to contemporary Putumayenses, not as isolated subjects or individuals but as interrelated embodiments of this relation.

In the first chapter, I sought to depict the ways in which the hegemonic discourses and practices of the state have been crafted and embedded *in* time and *through* space. In this manner, and especially through the figure of the nineteenth-century *criollo*, I tried to look at one end of the wide spectrum of characters comprising such relations. In this chapter I would like to consider the other end of this spectrum by focusing on the figure of the contemporary *desplazado* (displaced person).[1] Specifically, I seek to describe the turbulent resettlement process of a displaced community illegally occupying a segment of the road-project's area on the outskirts of Mocoa. Through an ethnographic description of the origins and

Frontier Road: Power, History, and the Everyday State in the Colombian Amazon,
First Edition. Simón Uribe.
© 2017 John Wiley & Sons Ltd. Published 2017 by John Wiley & Sons Ltd.

development of this process, I would like to lay emphasis on the multiple practices, strategies and actions through which the displaced struggle to overcome their condition of 'rightlessness', as well as to reflect on both the potential and limits inherent in this struggle. In doing so, I will draw attention to how the relationship of inclusive exclusion manifests itself in the daily lives of the displaced peoples, as well as the different ways through which they make sense of and challenge it.

The displaced

The number of internally displaced people in Colombia as a result of the ongoing armed conflict was estimated in 2015 at 6.9 million, the largest population of this type in the world (UNHCR 2015). These numbers, accounting for roughly 14% of the country's population, give an idea of the dramatic effects of decades of violent conflict. Forced displacement constitutes a social and human tragedy that defies easy analysis and is difficult to grasp through figures alone. Uprooting, loss of residence, severing of community and family ties, marginality and stigmatization, are just some of the conditions of forced displacement that are difficult to describe, let alone measure.[2]

At the centre of the political and academic debate concerning displaced people is the issue of rights. This issue stems partly from the simple fact that displacement always poses a serious threat to, or implies a loss of, basic rights such as work, shelter, health and security. In theory, the government has the constitutional obligation to restore and protect such rights, and Colombia is certainly known for having one of the most advanced legislative frameworks in the world in relation to internally displaced populations. However, the wide gap between legislation and practice is often cited in the literature as one of the main problems accounting for the displaced peoples' 'rightless' condition (CODHES 2014; Durán *et al.* 2007; Easterday 2008; Merteens and Zambrano 2010; Segura 2002). This situation, on the other hand, has been publicly acknowledged by the country's Constitutional Court, which at various occasions has declared as 'unconstitutional' the state's prolonged inability or unwillingness to guarantee the rights of the displaced population (Corte Constitucional 2004, 2009).

Although the subject of rights has brought attention to the highly vulnerable situation of the displaced, less attention has been given to the ways in which this issue transcends the particular condition of forced displacement. Daniel Pécaut (2000) has discussed this specific subject by analysing the phenomena of forced displacement in the long history of violence in the country. Drawing on Arendt's account of European refugees in the context of fascism and World War II, Pécaut reflects on the case of the Colombian displaced population around the three dimensions of displacement identified by Arendt: the loss of 'residence', implying not only loss of domicile and property but 'social texture' (e.g. community and family ties); displacement not as a consequence of individual or collective political or ideological standings but accidental or 'natural' reasons (e.g. belonging to or having being born in the 'wrong' class, race or place); and loss of political and civil rights resulting from a condition of 'stateless-ness' (loss of membership to a national political community).[3]

According to Pécaut, the Colombian internally displaced largely fits within Arendt's threefold condition of the European refugee. However, he emphasizes that the former, though 'rightless', cannot be considered a 'stateless' person in the strict sense of the term or the lack of membership to a nation state. Still, he suggests that if this membership be regarded in effective rather than just formal terms, then the Colombian displaced resemble a 'stateless' person. This distinction, otherwise expressed as the gap between nominal and actual citizenship, is crucial in Pécaut's argument, as it is here that the 'rightless' status of the displaced is revealed not as a temporary condition related to the situation of displacement, but as a permanent reality inscribed in the country's long history of violence. Hence Pécaut's statement that, in Colombia, displacement 'is not simply a life episode, but the living out of a near permanent social condition' (Pécaut 2000, p.93).

Pécaut's analysis is highly relevant to the discussion on forced displacement, especially as it shifts the debate from the *conjunctural* 'rightless' status of the displaced person to the *structural* relationship between state and society in a context of prolonged political and social conflict (see also Sanford 2004). However, the author's approach to this relationship is fundamentally different to the one I have taken in this book. Pécaut's argument, well-established within the tradition of violence studies, broadly maintains that this relation has been marked by a 'precarious' state that has historically 'coexisted' with private or extra-legal powers, or whose power vacuum in large areas of the country has been 'filled' by the latter.[4] Thus, for Pécaut, the real tragedy of the displaced is not that of a people rendered 'rightless' by the violent act of displacement, but that of a large population inhabiting a 'fragmented territory' whose membership or connection to the state has been in many respects 'a fiction in the face of the dominance of personal networks of power' (Pécaut 2000, p.94).

Like Pécaut, I think it is essential to recognize that the 'rightless' condition of the displaced transcends the contingent situation of forced displacement. Unlike him, however, I don't think this condition can be explained in terms of a situation of *exclusion* from the state, no matter if this exclusion is expressed in geographical, political or symbolic terms. Rather, I would suggest that the tragedy of the displaced, a tragedy that resembles or epitomizes in many respects the condition of the frontier, is not the tragedy of those excluded *from* the state but that of those whose relation *with* the state has been one of inclusive exclusion. The point I want to emphasize here is that this distinction between *exclusion* and *inclusion by means of exclusion* is essential in understanding the relationship between displaced peoples and the state in Colombia.

In addressing this point, it is useful to draw upon Partha Chatterjee's (2004) notion of 'political society', a concept he uses to explore and question the relations between state and society in the postcolonial world. Although this concept of 'political society' is not novel and has been used in other contexts, most notably perhaps in Gramsci's understanding of hegemony (Gramsci 1971, pp.206–276), Chatterjee recreates it through his critique of universal Western notions of the state and civil society. These notions, recalls Chatterjee, were grounded in a modern ideal of democracy, where the affiliation to the state was assumed as a universal guarantee of citizenship or a process that automatically turned individuals set apart by race or class into equal citizens, bearers of rights. In most of the world, however, and particularly outside the West, this ideal seldom matches the actual nature of the state and democracy, since the domain of civil society is confined to a fairly small segment of the population. The vast majority of people, in contrast, can hardly be considered 'proper members of civil society' or 'right-bearing citizens in the sense imagined by the constitution'; and yet, adds Chatterjee, 'it is not as though they are outside the reach of the state or even excluded from the domain of politics' (Chatterjee 2004, p.38). Thus, what actually differentiates these 'populations' from the small minority of 'citizens' is not their condition of exclusion or inclusion *from* the state, but the way in which they *relate* to it: a relation that is defined not by the conventional rules of democracy, as in the case of 'real' citizens, but by the messy and often conflictive terrain of politics.

This particular relation, which ultimately defines the essence of 'political society', has two main intertwined facets or dimensions. The first is about how governments deal with or aim to control and regulate those who under the eyes of the state are seen as subjects or populations rather than citizens. This facet encompasses the sphere of governmentality. The second facet is about what Chatterjee calls 'the politics of the governed', and has to do with the infinite array of informal practices through which these same subjects or populations claim their status as citizens. The illegal or paralegal character of such practices is of great relevance within the meaning of 'political society', for it is this character that unveils and exposes the fictitious nature of universal citizenship under the constitution of the modern state. But the critical point of 'political society' is not, Chatterjee reminds us, how it exposes this fiction but the fact that the fiction 'must be recognised and dealt with' (Chatterjee 2004, p.74). In other words, as long as the utopic condition of universal citizenship has to be maintained, those practices, though existing outside the formal rules of 'civil society', acquire a *legitimate* or widely accepted character in 'political society'.

These facets of 'political society' – facets that as Chatterjee argues constitute not deviations of modernity but one of their most distinctive features in most of the world – are, I think, crucial in comprehending the everyday realities and politics of displaced peoples both in the particular case addressed here and in Colombia in general. The first facet is highly illustrative of the process through which the displaced have entered the realm of official discourse and have been constructed as a specific object of government intervention. This is a subject that has been well addressed in the literature (see, particularly, Aparicio 2005), and which I have already discussed in the broader context of state making practices in the frontier. Here, my primary concern is with the second facet, for two main reasons. Firstly, I consider it provides a suitable theoretical and ethnographic framework to discuss and reflect on the sort of strategies, characters, events and conflicts around the resettlement process analysed here, as well as the role played by the road within it. Secondly, I think it allows for reflection on the possibilities, limits and challenges faced by those who, like the displaced peoples, inhabit a frontier condition and, as such, have been historically and spatially assimilated to the order of the state through a relation of inclusive exclusion.

The making of a community

The story of Guaduales is in many ways the story of one of the many displaced settlements that originated as a result of the war in the country's rural areas. It goes back to the early 2000s, when the escalation of the conflict between the army, the paramilitary and the guerrilla, together with the increase of aerial fumigation of *coca* crops in departments such as Putumayo, Caquetá and Guaviare, massively increased the exodus of peasant families to urban centres both within the region and to other parts of the country. In the Putumayo, in the year 2000 alone 1,397 displaced persons were reported to have arrived in Mocoa, which rapidly became the main recipient of the displaced population in this department, to the point that three years later it was estimated that the displaced comprised more than one-third of the town's total population (Sánchez 2007, pp.89–90). Although most of these displaced arrived initially to family or friend's homes or rented houses in marginal neighbourhoods of the city, many moved to already existing informal settlements that began to spread on the outskirts of town.

Guaduales was one of the first of these settlements, commonly known as 'invasiones' (invasions) due to their illegal character or the fact that they took place spontaneously on privately or publicly owned lands. Its founders or initial residents were about 50 peasant families from different parts of the region, most of who had met in the course of the bureaucratic pilgrimage through government offices and humanitarian agencies. The majority of these 'pioneer' settlers had left the place by the time I first visited Guaduales, having moved to other parts of town and the Putumayo or, though these were very rare exceptions, had returned to their lands. However, those who remained recounted to me the conflictive and difficult beginnings of the settlement.

It all started in October 2000. After weeks of search and deliberation, it was agreed that the best spot was a strip of pasture land along a municipal road going to Guaduales, one of the town's *veredas* and the name by which the future settlement eventually came to be known. Situated about two kilometres northeast of the town centre, the place seemed ideal in terms of access and proximity to basic services, government offices and jobs. However, due to its location both on public property (a road) and private lands (a farm), conflicts with the local authorities were anticipated. Thus, and notwithstanding their precarious means and resources, action was swift and as planned as circumstances allowed. As the general consensus was that numbers would play an important role in the initial struggles, the decision was made that everyone would squat at once, setting up a *cambuche* or improvised tent on the spot.

People recall the first days of Guaduales as the toughest in the history of the settlement. 'Stacked like animals' was an expression I often heard from the early settlers when describing the overcrowded situation of more than 150 persons, including men, women, elderly and children, all living under the same roof. A provisional board was created to deal with the most urgent issues such as food provision, cleaning and security, delegating tasks to every member of the settlement, regardless of age or gender. The most important question, however, was the survival of the camp while they reached a formal agreement with the township acknowledging their condition of displacement and hence the legitimate, if not legal, existence of the settlement until a permanent solution was negotiated. This agreement was eventually reached, yet only after fierce confrontations with local authorities. During the first three weeks, anti-riot police attempted to evict the residents and empty the place twice by firing tear gas, to which Guaduales' people responded with machetes, sticks, stones, and any other weapon that came to hand.

The turning point of the struggle came during the third eviction raid, at the end of the first month. Ramiro,[5] a mechanic from Puerto Guzmán (Putumayo) and one of Guaduales' early leaders, proudly related this landmark event in the history of the settlement:

> We knew they were coming on that day. So we sent for the garbage truck saying that there was a huge pile of trash here. And when it came we were waiting with clubs and machetes. We kidnapped the truck and its driver, laid it across the road and put cardboard underneath. I had a small motorbike and we emptied the fuel tank and poured gasoline over [the cardboard]. And when the police came we told them we would set the truck on fire if they moved a step forward. We put the children in front, then the elderly, then the women, and we stayed at the rear; all with wet wipes and bottles of water to protect us from the tear gases. We stayed there until the Mayor showed up – because he had said before that he didn't want anything to do with us, the displaced – and signed an agreement allowing us to stay there until we were resettled.

If the garbage truck episode represented an important victory for Guaduales' people, it marked just the beginning of a long process that continued for much longer than initially expected. Following this event, the next and most urgent step was for those included in the agreement, 56 families that were officially certified as 'displaced', to build temporary homes while the township sought funding for a 'formal' housing project. They were allowed to build their houses along a thin strip of land on both sides of the road, and plots measuring four metres in front by six metres deep were allocated internally by the board.

At first, houses were to a large extent family versions of the initial *cambuche*, made of black plastic sheets, cardboard and wooden poles. Yet, as time went by and the promise of new houses remained largely on paper, the settlement gradually began to look less like a camp and more like a neighbourhood of precarious but permanent dwellings. Plastic roofs were gradually replaced by zinc tiles and those who could afford it covered the dirt floors with cement or concrete. Cardboard on the walls gave place to wooden planks, a highly valued good for the displaced, to the point that many people in Guaduales made long journeys back home just to strip them from their abandoned houses or what remained of them.

The attainment of basic services implied long struggles and numerous collective strategies. Since from the point of view of local authorities Guaduales could not be regarded as a formal settlement and it never ceased to be perceived as an invasion, basic infrastructure was never 'officially' delivered. Water was obtained from a spring nearby and brought by a hose, and the supply was managed by one of the settlers who charged every household a monthly fee of 2,000 pesos (US$ .70 approx.).

This situation often gave rise to disputes and quarrels, some families refusing to pay and accusing the 'manager' of monopolizing and taking advantage of a resource they considered free, while others opted to install their own private hoses.

Electricity constituted another chapter. Initially, it was obtained by gathering and joining several pieces of old wire, making a long cable that was 'hung' from an electricity post located at the southern end of the settlement. This strategy, popularly known as 'piracy', did not last long due to the overload of users and conflicts with the local power company, causing constant cuts and short circuits. The company, tired of removing the 'pirate' cable just to find a new one set up overnight, agreed to provide legal power supply and installed an electricity metre for the whole settlement. Guaduales' people recall that when the first bill arrived, the sum was so large that no one even worried about paying or thinking how it should be shared. Days later, when one of the company's employees came to cut off the supply, he found himself on top of an electricity pole surrounded by a crowd of outraged neighbours complaining about the unfair bill and the company's poor service. Eventually, and after being offered a blanket with the implication that he would be obliged to sleep on top of the pole, the hapless technician had no choice but to reconnect the service. The conflict continued until a new agreement was reached with the company, who now installed an independent metre for every four or five houses, making it easier for neighbours to share the costs and control energy consumption. Though this arrangement did not solve the conflicts regarding bills and some continued to 'pirate' the service, it eventually became the prevailing system in the settlement.

***

In April 2010, when I began to visit Guaduales regularly and almost ten years after its birth, the settlement looked like a neighbourhood long since incorporated into the landscape of Mocoa (Figure 6.1). A *colectivo* (small bus) ran every half an hour to and from the town centre, where people travelled daily for work, shopping and errands. Most young people and adults had jobs in town – women mostly as housemaids and men as construction workers – while children attended one of the public schools nearby or spent the day in the Bienestar Familiar, the government's family welfare institution. The place, on the other hand, looked busy all the time. The road not only had a constant traffic of motorbikes and cabs going to the nearby *vereda* but also served as social space and playground for children. Some houses, around 15 out of 70, had *tiendas* (small grocery stores) frequented mostly by the same residents as they usually sold *fiado* (on credit) and in smaller quantities than the town's

Figure 6.1 Guaduales, April 2010.

shops. It was also common to see government and NGO workers around almost daily carrying out all sorts of surveys and campaigns, from vaccination to family planning and birth control.

Guaduales' liveliness, however, should not be interpreted as being indicative of a stable or consolidated neighbourhood and, in many ways, the place continued to exist as the marginal settlement it was in its early days. As previously noted, many of the original dwellers had left, due mainly to the same reasons or problems affecting other displaced settlements: insecure tenure, poor living conditions, lack of or precarious job opportunities or simply the social stigma of living in an 'invasion' neighbourhood.[6] The settlement, however, had not depopulated but actually grown over the years. New residents arrived regularly, usually renting or buying the homes of those who had left or, in some cases, building new shacks in the few available spots that remained. Yet, and for its most part, Guaduales' residents constituted a floating population for whom the place was more a transit location than a permanent home.

Under such conditions, the idea of community seemed difficult to conceive in a place like Guaduales. This had to do not only with the unstable character of the settlement, but also with the fact that most of its residents appeared to have little in common other than the condition of being displaced. Doña María, an Afro-descendant woman displaced from the lower Putumayo and one of the original settlers, summarized well this condition by describing Guaduales' inhabitants as 'a mix of everything: Indians, Blacks, Whites, everything'; and yet, at the same time she emphasized how despite class or racial differences and the fleeting character of the settlement, a 'sense of community' had developed through time.

During the months that I followed Guaduales' events and everyday conflicts, I corroborated that despite the unstable character of the settlement this sense of community did not only exist but was relatively solid. Furthermore, two related elements distinguished and sustained this sense of community. Firstly, and in contrast to what one might find in a conventional or 'formal' neighbourhood, it had been born and persisted in Guaduales not as the result of a long and continuous process of daily coexistence and interaction, but out of necessity. As mentioned, not only from the perspective of the local authority but to a large extent of Mocoa's citizens, Guaduales was always seen as an 'illegal' and hence undesirable space. For Guaduales' people, this situation meant having to bear not only the burden of stigma, but dealing with the harsh daily realities of informality, the most oppressive and challenging of which constituted the omnipresent threat of eviction. Beyond this threat, vital needs such as water, electricity and other basic services had to be secured, thus demanding collective action and a certain minimum level of political organization within the neighbourhood. Community, in other words, was not a matter of choice but of survival.

However, what underpinned the resilience of this community was not so much the challenges it faced daily, as its members' *status* of displaced. Being labelled or categorized as displaced persons, inevitably implied belonging to a population subjected to a series of homogenizing, often denigrating practices. However, at the same time this label can also act as a rhetorical and political vehicle through which this condition can be challenged. This was evident in the case of Guaduales, where the condition of displacement not only marginalized but also empowered, most significantly granting *political* legitimacy to certain claims and practices otherwise deemed illegal or unlawful under the town and state laws. Put differently, and going back to Chatterjee's notion of political society, the status of displacement became in Guaduales the identity feature that '[gave] *to the empirical form of a population group the moral attributes of a community*' (Chatterjee 2004, p.57, emphasis in original). The potential and limits of this transition towards a community, on the other hand, as I will describe in the remaining sections of the chapter, would only be fully exposed in Guaduales' long and conflictive resettlement.

The struggle for resettlement

The community organization process in Guaduales began with the settlement itself. As said, the original residents set up a provisional board to meet the community's most pressing needs, and above all to face the early eviction threats and handle the negotiations with the township. Although this board continued to exist for some time, it eventually gave place to a formal or 'legal' organization named 'Asociación de Desplazados del Putumayo' [Association of Displaced Persons of the Putumayo] (ASODESPU), formally established in 2003 and which survives to the present. This transition was not so much the result of the community's initiative, as of a legal requirement that demands community organizations to have a legal personality in order to channel their claims and proposals through state institutions. ASODESPU was initially formed by Guaduales' first 56 families, but in the following years it grew significantly as new people moved in and others living in other parts of town joined. By 2010, the association had about 200 families affiliated and though only a few were active members, it played an important role in dealing with the community's internal affairs as well as making visible its problems with government authorities.

However, ASODESPU's role was not confined to its strictly 'legal' functions or attributes. An equally or even more significant characteristic of it was essentially political. This dimension, otherwise common to the approximately 20 organizations of displaced persons that existed in Mocoa at the time, had to do, at least in part, with the precious value of the vote to both the politician and the displaced. To the former, and particularly those aspiring to local posts such as mayor or councillor, the vote of a displaced community can be decisive, especially considering the relatively small population of a town like Mocoa. As for the latter, the vote is not so much conceived as an expression of political will, as a commodity through which concrete benefits can be obtained for the individual and the community as a whole. Some of those benefits are long-term electoral promises recorded in written agreements (e.g. housing projects, delivery of public services or other basic infrastructure), most of which are only partially or never fulfilled. Others are immediate or gathered 'in advance' during the electoral campaigns, and range from food baskets and 'free services' (e.g. haircuts, medical consultation, legal advice) to free meals and even 'cash for votes' on election days.

The ability to make the most out of political opportunities both during and after electoral campaigns constitutes one of the essential roles of displaced persons organizations, one where the 'political' appears as a space where citizenship is not so much exercised as bargained on an almost daily basis. In the particular case of Guaduales, this political dimension is also crucial in understanding the different moments and events related to its resettlement process, a story that I gathered in fragments from some of the leaders who played central roles in the different stages of the process.

The first of them was Ramiro, whom I first met in one of ASODESPU's meetings during 2010. He was known for his confrontational and outspoken character, constantly generating debate and controversy. Ramiro no longer lived in Guaduales, having returned to Puerto Guzmán in 2008, where he ran a small motorcycle repair shop. He was one of the founders of 'Familias Unidas de Mocoa' [United Families of Mocoa], a smaller association organized by families that were initially rejected by ASODESPU, but eventually merged with it. According to him, the first attempt to resettle came from Familias Unidas, which on its own initiative decided to purchase a small plot of land near Guaduales. The land measured two hectares and was bought for 12 million pesos (US$ 4,000 approx.), a sum that was divided among the association's 70 members. The expectations were such that many people from ASODESPU joined for a second phase, and raffles and bazaars were organized to collect money for future settlement plans. However, and even though the township agreed to fund the project, this initiative was frustrated when they were denied the environmental permit required for construction on the argument that the plot was located in a rural area not zoned for urban development. The disillusioned owners had so far been unable to agree on what to do with the land, and so, lamented Ramiro, 'it was left abandoned and *criando monte* [overgrown]'.

The second initiative came from ASODESPU and began during the 2003 local electoral campaign. It originated during a meeting with Elver Cerón, one of the candidates for Mayor, who committed himself to implement a resettlement project if he was elected. The event took place at a cockfighting ring near Guaduales and Ramiro himself drafted a 10-point document in which the candidate promised, amongst other things, to purchase a land plot, obtain funds to develop a housing project and finance productive projects for the community.

Cerón was elected for the period 2004–2007, according to Ramiro thanks largely to the massive support he got from ASODESPU's members. However, most of his electoral promises included in the agreement never materialized or came too late. A first housing project, named Villa Rosa I and benefiting an initial group of 78 families, was approved in December 2005, at the end of his second year of office. Yet, and even though more than half of the project's total cost was secured through a housing subsidy from the national government, it remained stalled for more than two years. The reason for this long delay was mostly due to the fact that the land where the project was to be developed was not acquired until October 2007, less than three months before Cerón's term came to an end. Finally, during his last month, the Mayor approved a project for a second housing phase, Villa Rosa II, for 61 new families, a move that many in Guaduales regarded as a last minute strategy to demonstrate the good performance of his administration.

The construction works of Villa Rosa I began in March 2008 with the assistance of the beneficiaries, who as stipulated in the project's general guidelines, were to provide the 'unskilled labour' for the building of houses (Banco Agrario n.d.). Ramiro recalled the first months of construction as particularly arduous, starting with the issue that the project's land, a 62 hectares farm located in a peripheral area of Mocoa, did not have any access roads and thus the construction materials had to be brought in on the back of mules or by the beneficiaries themselves. To make things even more dramatic, the start of works coincided with the beginning of the rainy season, making the transport of materials and the construction itself a strenuous and sometimes impossible task.

It was in that context that the road entered the picture. During 2008, most of the technical studies related to the *Variante* San Francisco-Mocoa were completed and Guaduales was identified as one of the project's major concerns (INCOPLAN 2008, vol.3, pp.195–197). The settlement happened to be located in the future road's right of way and the project's starting point or 'kilometre zero'. And though all the assessments made clear that resettlement was beyond the project's scope and was a matter for the municipality, this situation gave Guaduales a notoriety it had not enjoyed before. The project's loan terms, for instance, set as the first condition for the initiation of works that the right of way should first be released (IADB 2009a), hence putting pressure on the township and the Mayor in particular, as well as involving other institutions in the process. One of them was the Invías, which had committed to build the access road from the town to the new settlement, thus solving the urgent issue of access.

For Guaduales' residents this new situation was unexpected as they had occupied the area before the final layout of the road was decided, but would become the bone of contention in the struggle for resettlement. In contrast to most Putumayenses, for whom the future road embodied a long awaited promise of 'progress' and 'development', for Guaduales' people it meant a tangible political asset, rather like the vote but perhaps more valuable considering the relevance and magnitude of the project. The question, then, became how to 'capitalize' on this asset given the many conflicts and events that accompanied this new chapter in the history of Guaduales.

A central character in how events would now be determined was Rubén, a displaced person from the neighbouring department of Caquetá and president of ASODESPU's board between 2007 and 2009. Rubén, more than any other leader I met, emphasized constantly the importance of the displaced leaders' *gestión*, a staple and yet complex term within Colombia's political and bureaucratic jargon that is best translated as the ability to get things done. Rubén himself tirelessly praised his own *gestión* throughout the process, something he credited to his 37 years of experience (he was 59 when I first met him) as community leader and his extensive network of contacts. Thus, he largely attributed to his personal *gestión* most of Villa Rosa's progress and relevant events: the agreement signed in June 2008 between ASODESPU and Mario Narváez, the new Mayor, where the latter committed to deliver the houses of Villa Rosa I and II by March 2009; the allotment of houses for the beneficiaries; the procurement of food baskets for those working on the construction; and the obtaining of resources to build a children's home for the new settlement.

Despite Villa Rosa's progress under Rubén's term as president and the growing pressure of the road project to speed up the resettlement process, the new settlement did not move ahead as anticipated. When Narváez's deadline arrived, the houses of Villa Rosa I were half built and the construction of Villa Rosa II had barely started. However, and despite ASODESPU's pressure to mobilize against the Mayor, Rubén refused to support any confrontation outside institutional channels. This situation came to a head in November 2009, six months after Rubén had finished his term and seven months after the Mayor's promise to finish the project. Eustasio, ASODESPU's new president, narrated the event:

> We were fed up with the negligence of the township. So we decided to carry out a coup so the Mayor would hear us and take us seriously at last. So on the tenth of November, around 3 am, we took over the township building. All the community was present. And that day all the authorities showed up: the municipal ombudsman, the public prosecutor, the police, the township officers. And new agreements were signed: the finishing of houses, water and electricity services, the construction of streets, etcetera. I mean those agreements were not new but were given a new deadline.

The take-over of the township constituted in many ways a natural and foreseeable response to the municipality's continuous disregard for Guaduales' countless claims and petitions regarding Villa Rosa. However, this event also reflected a change of strategy on the part of the displaced, a shift that responded in part to the arrival of the new president. Eustasio was a coffee farmer who had also served as professional soldier in a counter-guerrilla battalion for 12 years. Since his arrival in Guaduales, in 2016, he had been actively involved with ASODESPU, and during Rubén's term had been in charge of the association's committee of works. However, and differently to Rubén, Eustasio's notion of *gestión* was grounded not so much on political negotiation and networking but on public confrontation. Thus, it was a salient feature of his speech to stress that the community should 'take matters into their own hands' or 'take action' if their demands were not met through the regular channels, or if politicians failed to comply with their promises.

If the township 'coup' reflected ASODESPU's new politics under Eustasio, the significance of this event can hardly be reduced to a change or transition on the organization's leadership. Rather, I think, this significance is to be found in the longer history of Guaduales, and especially what it revealed of both the changes and continuities in the community's struggle. It represented a substantial change, for this event would have been unthinkable in the early days of the settlement, where the demands of the displaced were met with tear gas and clubs. And while this change was undoubtedly related to how the phenomena of internal displacement had during the past years gained relevance in the government's agenda, the event also revealed an organized community capable of mobilizing in order to make its voice heard. Yet, the fact that in order to be heard and 'taken seriously' such an extreme and desperate measure had to be taken, also showed how little the *nature* of the relation between the displaced and the state had evolved over time. In this sense, the township event did not represent a step forward in Guaduales' struggle for basic rights, but confirmed that the attainment of such rights continued to be mediated primarily by the conflictive and unstable terrain of political confrontation and compromise.

This oscillation between change and progression on the one hand and continuity or persistence on the other, constitutes a central aspect of Guaduales' resettlement history. This aspect, moreover, could not be captured only through the leaders' memories of landmark events, but from Guaduales' people's everyday accounts and anecdotes about the settlement, all of which told the same tale: moments of advancement followed by stagnation and setback, small victories and breakdowns, people coming and leaving, fits and starts, times of expectation and dismay. Villa Rosa, in this sense, did not constitute a break but a continuation of this fluctuating movement, with the difference that the drama was now transposed to a new setting. However, as I shall describe next, the intensity of events and conflicts that accompanied the beginnings of the new settlement made it possible to observe those changes and continuities in a compressed time frame.

Villa Rosa

The most telling fact about Villa Rosa is perhaps that it was never officially inaugurated: no cutting of tape ceremony, no speeches, no press release. Guaduales' people do not even recall a specific date. According to Doña María, in early March 2010 the presidents of the association boards (ASODESPU was now internally divided into Villa Rosa I and II) handed over the house keys to the beneficiaries. 'Rumours were that we should move in at once or otherwise the houses would be re-allotted to other people' she said, while noting that the event coincided with an announcement from the township notifying them that Guaduales should be emptied straightaway. However, only a very few complied with the order. Most put in some stuff or began to visit the place regularly to 'mark territory' but refused to move, for the place, they said, was 'unliveable', a word that accurately captured the precarious conditions of the new settlement.

To begin with, even by the standards of social housing projects, the house size was extremely small (27 square metres in Villa Rosa I and 37 in Villa Rosa II), especially considering that they were conceived as homes for families averaging five members. Certainly, if regarded in terms of their 'features' – two-room rectangular structures of bare cement brick and dirt floors – they could be better described as brick shelters rather than actual houses. They had no bathroom or kitchen, no electricity, and no water supply. They came with septic tanks but most of them were useless as they had been dug on clay soils with low absorption capacity or, even worse, were positioned above the house level, a situation that some beneficiaries sarcastically remarked implied that 'the constructors probably wanted us to shit uphill'. In addition, some of the houses were built on unstable ground, so they already had cracks and fissures on the walls. On top of all this and to complete the new settlement's grim diagnosis, no internal street network was built, so most of the houses had to be accessed, or were connected, by muddy pathways (Figure 6.2).

In early April 2010, when I visited Villa Rosa for the first time, its condition had barely improved, and the conflict between the township, Guaduales' community and the road project was at boiling point. The Mayor's time limit – the one agreed upon during the township take-over on November 2009 – to deliver the project in full was a few weeks ahead, but at this point it was clear that this deadline, like the previous ones, would not be met on time or even in the near term. Invías, meanwhile, urged to release the right of way to allow the start of the road works in the Mocoa section, pressed the township to carry out the resettlement without further delay. The Mayor, for his part, kept sending eviction notices to Guaduales whilst at the same time assuring ASODESPU's leaders that he was making the respective *gestión* to solve Villa Rosa's problems. Yet Guaduales' people refused to leave, and responded to the threats by showing up at the Invías offices and other local authorities with the several unfulfilled agreements from the actual

Figure 6.2   Section of Villa Rosa, May 2010.

and previous Mayor. Eustasio even spoke with some regularity on local radio to counter the public perception, fuelled by the Mayor himself he thought, that Guaduales was unwilling to resettle despite the housing 'solution' they had been provided with.

In this vicious circle of mutual accusations and claims, the most obvious truth about Villa Rosa was exposed – that the houses' precariousness and unfinished state was nothing but a reflection of the uncertain political arrangements and promises on which they were grounded. The first and most critical issue was that Villa Rosa never had a housing design plan *per se*. According to the township secretary of projects under Elver Cerón, the Mayor 'showed up one day with a list of IDs to be presented as beneficiaries for housing subsidies in the Banco Agrario'.[7] Cerón, he added, actually requested a design of the housing plan, but at the end it was never carried out since the construction of houses was contracted several months prior to the acquisition of the land (Personal interview, 12th May 2010). As a consequence of this incongruity, the actual 'housing plan' ended up being basically a random division of the land into individual plots measuring an average 20×30 metres (600 square metres). Another equally striking fact related to the houses themselves. In Villa Rosa I, the most controversial case, the houses' plan showed structures of 35 square metres and included one bathroom and a small exterior kitchen. Nevertheless, the project's guidelines approved by the Banco Agrario stated that the subsidy was for 'housing improvement' instead of construction (Banco Agrario n.d.). Although this detail was never clarified and ended up adding to Villa Rosa's list of 'unresolved' questions, the general consensus, including that of the current Mayor, was that the small amount of the subsidy was the reason why the bathroom and kitchen were never built.

While the current Mayor publicly blamed Villa Rosa's many flaws on the previous administration, Elver Cerón, the political architect of the project, obstinately defended his involvement in the matter. When I interviewed him to talk about his role in Villa Rosa, his answers were for the most part an exaltation of his own *gestión*. So, when I brought up the housing subsidy issue he avoided the question by stressing that at the time the available subsidies were 'too small' so he had to 'echar mano de lo que había' [get a hold of what they had]. He emphasized the hardships he faced to find a piece of land that suited the 'dream' of a 'rural style' urbanization. This dream, he added, was shared with Guaduales' community, and one of the points agreed upon was that the settlement would not have streets; thus, and leaving no room for argument, he firmly stated: 'They were warned: this is a farm. And it will be provided with paths. But it will have no streets'. But the main problem of Villa Rosa according to the ex-Mayor lay not in its construction or defective design but on the 'lack of a strong hand' throughout the process: lack of a strong hand on the part of his prede-cessor, who allowed Guaduales' people to settle there in the first place; and lack of a strong hand on the part of his successor, too afraid of 'taking the matter in hand'. He claimed that those currently refusing to resettle were not 'real displaced' – for according to him these 'thanked god for having at least a house' – but opportunists looking to take advantage of the situation. Thus, and faithful to his strong hand approach, he stated his solution to the problem: 'It's a shame it is not in my hands. Then it would be like this: "take your new home, give me yours", then you have the old house demolished and no one will be there again'.

Cerón's views on Guaduales' displaced people are worth reproducing here not for the political rhetoric in which they are framed, but for the underlying perceptions in which the rhetoric is rooted. The displaced individual appears inexorably linked to a series of immutable realities he cannot escape and on which his assimilation to the spatial and social order of the city depends: he is 'rural' (as opposed to 'urban'), and as such he must be resettled in a 'rural style', a style which in the ex-Mayor's vision means a clearly segregated space from the town, whose main landmark is the absence of streets. On the other hand, the displaced are implicitly assumed here as 'rightless' subjects not on account of their *legal* status but because of their *natural* endowments. This is plainly evident in the emphasis on Guaduales' 'real' (as opposed to 'false') displaced person: the former – peasant, poor, dispossessed, homeless – are infallibly recognized by the fact that they are grateful and willing to move to their new homes, regardless of their uninhabitable condition; on the contrary, those conditioning their resettlement to the fulfilment of certain basic rights are by definition 'opportunists', a list in which the ex-Mayor included money lenders, traders and even 'well-off people' from town.

Perceptions of this sort were by no means confined to the politicians' rhetoric and during my time in the Putumayo I saw how they were reproduced in many other contexts. For instance, Doña María indignantly recalled once how a group of social workers and physiologists had come to Guaduales to teach the women how to 'inhabit' the new houses, and suggested that the 'space issue' was related to the displaced people's lack of family planning rather than flaws in the construction and design. Another example that had become a regular joke among Villa Rosa's beneficiaries was the explanation, brought up by the constructor, that the houses' unfinished state had to do with the fact that they had been originally conceived as 'self-construction' homes or as a base from which inhabitants could make sequential improvements.

Together, these perceptions could be seen as a 'language of stateness' (Hansen and Stepputat 2001, p.5) through which the inclusive-exclusive relation between the displaced and the state was daily enacted. However, it was also through this same language, through its systematic appropriation and subversion, that this relationship was constantly defied and challenged. Nowhere was this more evident than in the different 'encounters' between the community and the local authorities. These encounters were largely performing acts where some of the conflicts that surrounded the resettlement process were dramatically staged, while others were deliberately silenced or ignored. Furthermore, as performing acts, they confronted roles rather than individuals: the township officers as 'the authority', Guaduales' residents as 'the community' or 'the displaced', and so on.

An encounter of this kind took place in June 2010. At this stage, most of Villa Rosa's issues remained unsolved and the tensions between the community, the township and the other institutions involved in the project continued to rise. Conflicts regarding the new settlement were also affecting Guaduales' community from within. A few days earlier, at an ASODESPU's assembly, Rubén had quit the board amid a heated discussion in which he was accused of charging a fee to allot new plots in Villa Rosa. In addition, some families had recently moved to Villa Rosa and rented or even sold their old homes, breaking both the commitment with the township of dismantling the houses upon leaving, and the internal arrangement of remaining in Guaduales until the agreements with the township were fulfilled. It was in this troubled environment that the township and the Invías convened a meeting at Guaduales to address the different problems relating to the resettlement process.

The event was scheduled for the 17th June at 2:00 p.m., and despite the hour most of Guaduales' residents were present. The setting, arranged by the residents themselves in one of the few houses with a cement patio, consisted of a dining table covered with a white tablecloth and a few plastic chairs. These were reserved for the 'authorities', namely the Mayor, the local Director of Invías and the township Secretary of Planning, while the 'community' was to remain standing. At 2:00 p.m. the residents slowly began to gather at the spot, some bringing chairs and umbrellas to shelter from the sun. The first authority to arrive, around 2:15 p.m., was the Secretary of Planning, a young woman who rode a motorbike while carrying a thick folder and a USB memory stick hanging from her neck. The Director of Invías, easily recognizable by the institution's official 4x4 white Toyota, appeared at 2:30 p.m. At this point the place was quite crowded and the air was getting a bit tense, people starting to hear and spread rumours about the whereabouts of the Mayor. He finally showed up at 2:45 p.m. offering no excuses for his lateness and took the central chair, leaving the director of Invías on his right and the Secretary on the left. The latter, visibly relieved, quickly proceeded to read the agenda, a four-point document rigorously following the official protocol and consisting of interventions by the Mayor, the Director of Invías and the community – in that order – and, finally, the reading and signing of the meeting minutes.

Leaving aside the loaded visual language of this ritual – the gestures, the setting, the tacit hierarchies, the observed protocols – if one had to summarize the event in one word, the word would be *voluntad* (good will). The term was so overwhelmingly present throughout the meeting, that at some point it seemed as if it had become another actor on the stage with an agency of its own. It was first invoked by the Mayor, who devoted most of his speech to exonerate his administration from any blame regarding Villa Rosa. The main problem behind the delays and the current state of the settlement, he began arguing, was the rush and improvisation on the part of the previous administration to get the housing project approved, and the lack of resources to get it finished. Yet, he added how despite this obstacle he had always demonstrated *voluntad* to help the community (here he addressed the Invías officer as well), and by way of evidence he proceeded – with assistance of the Secretary – to enumerate the progress achieved lately: resources had been secured for the extension of houses in Villa Rosa I, the water supply network was under construction and he estimated it would be ready 'in a month or so', and the township council had recently allocated a budget to build the street network. Then, and to provide further proof of his *voluntad*, he recalled how in the previous

December the township had sent some 'little Christmas presents' to the people of Guaduales. Therefore, and as a way of retribution, he concluded, the township expected 'the community' to have patience and, above all, *voluntad* to resettle in a 'timely and orderly fashion'.

Upon this last remark expressions of indignation were heard among the crowd, but the Secretary called for 'order and respect for the agenda', and proceeded to give the floor to the Director of Invías. He began by highlighting the magnitude and significance of the road project, in his words 'aimed at replacing one of the most dangerous roads in the country and the world', and how there were 'growing concerns in *The Bank*' (the IADB, the project lender and an absent yet omnipresent authority everyone invoked as a sort of almighty power) and the Invías headquarters in Bogotá about the current situation in Guaduales. He then stressed that they were conscious of their 'difficult situation' and had tried to understand it, 'though', he emphasized, 'you who live it know better', and thus, he stated, they had sent sociologists and social workers on different occasions to ask how they were living and what was the situation of Guaduales. In doing so, he continued, they had also 'demonstrated *la voluntad de ayudar* (good will to help)', even though, he hastened to add, the Invías and *The Bank* were not responsible for the resettlement process. This good will he further illustrated by mentioning the different occasions in which Invías had allocated resources for the construction of access ways and bridges, and following this he declared that: 'we keep *la voluntad* to continue supporting you, but we need to get something in exchange, which is, in a very near future, to have the way released'. And thereupon he added, saying he was 'going to be clear' that they needed to know if this *voluntad* did or did not exist and, in such case, they would have to look for a 'different alternative', most likely a different route bypassing Guaduales. This, he concluded, was the 'message and the question' he brought from Bogotá and *The Bank*, where 'they', he again stressed, 'were very concerned about the situation of Guaduales'.

It was now 3:30 p.m. and the turn for the 'community' to speak. Eustasio took the floor and, following protocol greetings, began by stating that – he addressed the Director of Invías here – they had never said they did not want to resettle since they were the most affected by staying put; but 'they', he went on, now addressing the Mayor, 'had some agreements that had been signed at some meetings'. And then, addressing both the community and the authorities, he declared: 'We have *la voluntad*, I know the community has *la voluntad*. What we are asking from the township is the delivery of basic services: as displaced, as any human beings, we need those services'. Eustasio's claims were then repeated by some of the beneficiaries present, and each intervention was followed by applause from the 'community'. A man in the front row eagerly protested that the problem was not *la voluntad* but the endless delay in the delivery of services, and added that it was no one's secret that those already living there were forced to 'pirate' the electricity. A woman towards the back angrily asked if the water supply works the Mayor referred to were 'some chunks of pipe' that had been laying uninstalled in Villa Rosa for months and said they were currently 'collecting water from a ditch'. The Mayor replied that it was true there was no water but 'the solution' was already there and asked for patience while reassuring that the township had 'all *the good will* to help'.

Arguments of this kind continued for another ten minutes until, at around 4:00 p.m., the director of Invías asked for the floor again and, assuming a conciliatory tone, declared that the important thing was that all the parts had demonstrated *voluntad*. He said he would take this 'message' back to the Invías' offices in Bogotá and *The Bank*, and suggested scheduling another meeting within a maximum period of 20 days 'to make commitments and define time frames'. Even though faces of scepticism and resignation were everywhere to be seen among the crowd, the motion was unanimously approved and the Secretary hastily proceeded to read the minutes and have them signed by those in attendance. Upon finishing this the Invías' director and the Mayor, the first ones to sign, walked back to their respective vehicles and left, the residents slowly dispersed towards their homes, while Eustasio made arrangements with the Secretary to get a copy of the document. At 4:30 p.m. the setting was totally cleared and the place empty.

Despite such profuse demonstrations of *voluntad* displayed throughout the 2½ hours that the encounter lasted, no meeting took place within the agreed time frame – the next meeting took place in October, this time convened by Guaduales' surrounding neighbourhoods – and the conditions of the resettlement remained largely unaltered. Yet this encounter, as well as previous and future ones, was of crucial significance if considered not only in terms of the explicit commitments recorded, but of the implicit or tacit agreements reached. The recurrent invocation of *voluntad*, in this sense, acted not as a gesture void of meaning, but as a mutual rhetorical manoeuvre aimed at guaranteeing an equilibrium or *status quo* that at this stage benefited the 'authorities' and 'community' alike. For the township, and particularly the Mayor, this *status quo* meant another deferral to the commitments regarding Villa Rosa, most likely with the expectation that the 'problem' would be inherited – as ended up being the case – by the next politician in office. For Invías, responsible for overseeing that the resettlement was carried out according to *The Bank*'s policies (as was never the case) it meant keeping its role as a mediator or neutral actor while the process ran its slow course. And for Guaduales' residents, highly conscious that the road was their most valuable asset in the conflict with the township, it meant a time extension of their stay in the old settlement while their demands concerning the new one were gradually addressed. These would eventually be met, and yet only through the same sort of conflicts, crises and struggles so distinctive of political society.

On continuity and change

Amid the unstable *status quo* of sporadic political arrangements, renewed promises and endless daily conflicts, Villa Rosa progressed slowly. Part of this progress derived from the piecemeal interventions and projects of the various institutions and agencies directly or indirectly involved with the displaced population. Thus, the water supply problem was provisionally solved by a long hose donated by the Office of the United Nations High Commissioner for Refugees (ACNUR), funding for the kitchens and other house improvements was obtained through the International Organization for Migration (IOM), and the completion of the settlement's street network was funded with additional resources from the Invías. The other part came from the beneficiaries themselves, both through individual improvements to their homes and *mingas* (collective work sessions) during weekends and holidays (Figure 6.3).

Figure 6.3   *Minga* for the construction of the path. Villa Rosa, November 2010.

Despite such improvements, in December 2010 only a relatively few families (around 30 out of the original 139) were living in Villa Rosa, and seen from a day-to-day perspective the evolution of the new settlement seemed stalled. During the next two years, I returned for short visits in order to follow up its progress on a longer-term basis. In November 2011, the first impression upon arriving to the place was of an uneven construction landscape of many shapes and tones: some houses were radically transformed, the grey cement facades painted with bright colours or adorned with potted plants, the original areas expanded by the building of additional rooms, and the plots filled with grown crops and chicken coops; others showed ongoing construction work, with piles of sand, gravel, cement and other materials stacked in front of them; others were left unfinished and seemed abandoned or never occupied, with weeds growing on the walls and on top of the roofs.

The settlement exhibited some progress as well. The electricity supply had not yet been provided, the installation of the network in Villa Rosa I was almost completed (Villa Rosa II would take another year) and the township had promised connection to the service by Christmas. The water had finally been delivered and consisted of an external water tap for each house (for they had no plumbing installed), but people complained that it did not reach all the houses and that the pressure was too low. The streets had also been completed, although they were more a series of bulldozed paths than a proper street network, and some remained impassable during the rainy months.

Electoral banners displayed on the houses' walls and public spaces showed that 2011 had been an intense political year. Municipal elections had taken place in the previous October, and Elver Cerón had been elected as Mayor of Mocoa for the second time. According to Doña María, despite the distrust towards the politician in relation to the promises left unfulfilled during his first term, many in Villa Rosa and Guaduales had again voted for him, in the hope that 'he would come back and finish the job he had begun'. Yet, in contrast to 2003, no specific electoral agreements had been made with him or any other candidate, a situation partly reflecting the current weak state of Villa Rosa's associations. Months before, Eustasio had disappeared suddenly with no apparent reason and there was no trace of him, but the rumours were that before leaving he had been selling plots to non-displaced persons from Mocoa. Villa Rosa I had appointed a new president, but the association was now partly divided between those who were already living there and those who still refused to leave Guaduales (about half of the original beneficiaries), and the board complained of the low attendance at the meetings and *mingas*. Rubén, for his part, had been working since his withdrawal from Villa Rosa I's board on the development of a third urbanization phase called Villa Rosa III, a project for those residents from Guaduales excluded from the two initial projects. Most of these residents were officially classified as 'vulnerable population' or 'structural poor' rather than displaced people, and were commonly described by the latter as 'displaced by poverty'. Villa Rosa III had its own board and although Rubén was not officially part of it he was usually the one who allocated the plots and made the required *gestión* for the procurement of housing subsidies and other supports.

In addition to the three 'legal' housing projects, a new settlement known as 'El Paraíso' (Paradise) had appeared on the mountain located on the north side of Villa Rosa. Villa Rosa's residents considered this land part of their settlement, and said they had been told by the township that it was part of a Forest Reserve area where no constructions were allowed. No one in Villa Rosa seemed to know or relate to the new neighbours during the time I was present, and even though the area had been internally allocated in the form of farming plots as a sort of contingency measure to demonstrate possession over the land, the new neighbourhood was growing rapidly.

On my next trip to the Putumayo, in December 2012, Guaduales was gone. The last families' decision to leave had not been motivated by the continual eviction threats, or by the delivery of the township commitments regarding Villa Rosa, which were never completely fulfilled. Rather, the decision had been made after the road project and more specifically the Invías, in order to clear the project's right of way, had offered the remaining residents a cash incentive – purportedly for them to finish their houses – on the condition that they dismantled their current homes and cleared the intended route of the road. The last family had left the old settlement two months before and the few remaining vestiges of Guaduales were being rapidly covered by brush, to the point that it was difficult to imagine that a hamlet had ever existed there.

The new settlement, in contrast, looked fully inhabited and alive. Apart from the 139 houses of Villa Rosa I and II – now almost totally occupied – Villa Rosa III had 15 houses in different stages of construction. The children's home had been finally completed and was also being used as a community meeting place, and there were plans to build a small chapel with the support of Mocoa's parish. Another significant improvement achieved that year had been the establishment of a bus route to town, making the former 45-minute walk to the centre or the costly taxi service avoidable. Despite these developments, it was difficult to describe Villa Rosa as a stable neighbourhood and the general sensation, as previously with Guaduales, was of a place with people coming and leaving regularly. For instance, even though the original agreement with the township was that the beneficiaries should occupy the homes for an uninterrupted period of five years to be granted ownership, this regulation was rarely followed, and in practice it was common for families to leave temporarily or permanently and rent or even sell their houses to newcomers.

El Paraíso had expanded quickly and around 60 houses (compared to 20 from a year before) could be counted, the majority of which were *cambuches* similar to the ones found previously in Guaduales. Yet, and somewhat paradoxically, Villa Rosa's residents showed certain aversion towards the new settlement, and remarked with scorn that 'the place is called Paradise but looks like hell'. In the last year there had been some approaches, but the relationship between the two neighbourhoods remained mostly of mutual indifference, apart from a recent friction around the fact that El Paraíso was 'pirating' the water and energy from Villa Rosa. However, El Paraíso's main conflict was with the township, which insisted on the illegal character of the settlement, while its residents claimed their status of displaced persons and thus demanded a formal housing scheme.

* * *

Listening to what people from Villa Rosa and outside (township officers and NGO workers) commented about the sort of claims, conflicts and dynamics present in El Paraíso, it was impossible not to think of Guaduales a decade earlier. Both settlements seemed part of a same history, shaped not by a progressive linear sequence but rather a constant oscillation between change and continuity, a never-ending cycle where a frontier space gradually closed while other emerged.

There are many different ways to think of or account for both these two dimensions, some of which I have attempted to capture in describing some of the conflicts, events and characters comprising the long and troublesome transition from Guaduales to Villa Rosa. For instance, we could portray change by pointing at the most visible or tangible differences distinguishing the old and new settlements: the improvement of the houses' structures from cardboard, wood and plastic to concrete and cement, the transition from living under constant eviction threats to a situation of relatively secure tenure, the delivery of legal public services, or simply the reduction of stigma stemming from inhabiting a place called 'urbanización' [housing development] – no matter its precarious character – instead of 'invasion'. We could also think of change by considering other relevant and yet not so palpable developments and dynamics, without which the former, more visible changes would hardly be conceivable. Most significant, perhaps, is the process through which an uprooted and alien population group gradually acquired a sense of community vital in their struggle for rights and recognition. The wide array of tactics, strategies and manoeuvres encompassing and sustaining this struggle would, as I have tried to illustrate, be unthinkable without the persistence of this sense of community, despite the many obstacles, conflicts and contradictions to which it is continuously subjected.

I have approached and addressed the struggles of the displaced drawing on the notion of a political society, a concept which, as noted, aims to shed light on the lives of those whose relationship with the state is not mediated by the formal rules of democracy but the deeply unstable and uncertain realm of politics. This distinction is, as Chatterjee stresses, critical in understanding not only the potential and possibilities of such struggles but their own limits. I drew attention to these limits by emphasizing the sort of continuities or permanencies underlying Guaduales' resettlement process and, especially, the ways in which they stem from or reflect the fragile and messy political ground in which this process was rooted and sustained. This political ground, on the other hand, was not just confined to the ceaseless disputes and clashes between community and authorities, but the different conflicts intrinsic to both, for instance the political disputes between the township and the road project or the conflicting visions on the principle of *gestión* among Guaduales' leaders.

As in the issue of change, addressing continuity requires us to account not just for the visible or manifest. In particular, I have suggested how underneath the different policies and actions towards the displaced we find a series of deeply rooted perceptions, which together strongly mirror the inclusive-exclusive relationship to which the 'rightless' – yet not stateless – people are subjected to daily. In doing so, I have argued that the displaced personify in many ways the condition of the frontier: perpetually associated with a state of violence – even though it might be recognized as an *effect* of this same violence – and, as such, relegated to exist in social and spatial margins; in other words, included or assimilated into the order of the state through a series of exclusionary political, economic and social practices.

Nowhere in Guaduales' resettlement history was this relation so clearly exposed as in Villa Rosa: from the settlement's peripheral location to the tiny size and 'structural' design flaws of the houses, the unfinished and precarious state in which the project was delivered, and the ex-Mayor's statement that the 'real' displaced were those grateful to God for the housing solution they had been offered. Yet the crucial aspect of this relationship lies not only in its pervasive effects or manifestations, but in that it forces us to think of social and spatial constructs (state, frontier, community, and so forth) not as dichotomous realms but parts of a same ensemble. In the particular case of Guaduales, I described how the sense of community was inexorably tied to the perceptions and forms of exclusion to which its inhabitants were daily subjected. These perceptions and exclusions, in other words, speak not only of how the displaced have been produced as a governance category and in this way assimilated (spatially, politically and socially) to the order of the state, but how displaced people make sense of and respond to the physical and symbolic violence of this order.

Forced displacement constitutes a phenomenon that has to be understood in the larger context of Colombia's long history of conflict, whose analysis is beyond the scope of this book. In conceiving this phenomenon and in particular Guaduales' resettlement process in terms of the relationship between frontier and state, I have attempted to expose and question some of the discursive and material practices through which this violence is sustained, reproduced and contested. This is precisely why I think the contrasting image of Villa Rosa and El Paraíso is so revealing of this relationship – because it speaks of both the persistence and pervasiveness of this relationship and how frontier peoples struggle with it in everyday life.

Apart from Guaduales' story, there are many different instances, places and characters where this relationship manifests in the contemporary space of the frontier. I have described some of them in the second part of the book, together with the responses and struggles they elicit: the *trampoline of death*'s moral history and geography, the enduring disputes and pleas for 'decent roads', and the San Francisco-Mocoa road project land conflicts. These responses and struggles comprise different forms of political practice, and all pursue very specific interests, strategies and goals. There are many others beyond the context of the road, operating at different scales and exhibiting different degrees of organization and visibility: indigenous movements, peasant organizations, and civil society initiatives. Some of them have a long history in the region, whilst others relate to particular conjunctures and have an ephemeral character. Although it has not been the purpose of this book to build a typology of these forms of political practice, I think that their existence reflects in different ways the persistence of the frontier-state relationship, and especially the structural and infrastructural forms of violence that this relationship sustains. In the concluding chapter, I will further reflect on this relationship by referring to other events and situations that draw attention to the state's continuous dependence on the production and preservation of frontiers.

Notes

1   The Colombian law (Law 387 of 1997) defines the displaced person as: 'Any person who has been forced to migrate within the national territory, abandoning his place of residence or customary economic activities, because his life, physical integrity, personal freedom or safety have been violated or are directly threatened as a result of any of the following situations: internal armed conflict, civil tension and disturbances, general violence, massive Human Rights violations, infringement of International Humanitarian Law, or other circumstances arising from the foregoing situations that drastically disturb or could drastically disturb the public order'.

2   The literature on internal displacement in Colombia is very extensive. For a partial list of publications on the subject see http://www.acnur.org/t3/recursos/publicaciones/

3   For Arendt's original discussion on these dimensions of displacement see Arendt (1951, pp. 266–298).

4   For a general review of Pécaut's argument on the 'precarious state' see Bolívar (2003) and González, Bolívar and Vásquez (2003).

5   All the names of Guaduales' interviewees are pseudonyms.

6   See Sánchez (2007) for a detailed socio-economic analysis of the displaced settlements in Mocoa.

7   The Banco Agrario is the public institution through which the government house subsidies for Villa Rosa were channelled.

Conclusion
The condition of frontier

On 14th December 2005, during an official visit by the Brazilian president Lula da Silva, the then Colombian president Alvaro Uribe took his colleague to a room in the Presidential Palace where a portrait of Rafael Reyes hangs. Once there, he narrated to Lula the General's epic adventure in the Colombian and Brazilian Amazon in the 1870s, and commented how this 'visionary' and 'achieving' president was the precursor of the Initiative for the Integration of the Regional Infrastructure of South America (IIRSA). For, he informed Lula, Reyes had once 'put the saddle on his mule' and 'uprooting the forests' reached the plains of the Putumayo, and from there the Amazon River, the Atlantic Ocean and finally Rio de Janeiro.

The episode, narrated in a speech made by Uribe later that day during a lunch in Lula's honour, served to inform publicly about the important steps that Colombia had taken towards regional integration. The Colombian government, noted Uribe, had undertaken enormous efforts to fulfil Reyes' dream, and he was pleased to announce that the bid to build the San Francisco-Mocoa road, 'one of the Continent's major roads', was soon to be opened. 'It would be great Mr President!' Uribe effusively addressed Lula, 'that in the course of a few months we see this road project approved and its construction initiated, because this would be the beginning of a great dream' (Presidencia de la República 2005).

Despite such promising words, when Uribe's presidential term came to an end in 2010, not a single kilometre of the road had been built. The construction work eventually began in 2012, after many bureaucratic delays and conflicts. In the following four years, I travelled to the

Frontier Road: Power, History, and the Everyday State in the Colombian Amazon,
First Edition. Simón Uribe.
© 2017 John Wiley & Sons Ltd. Published 2017 by John Wiley & Sons Ltd.

Putumayo on various occasions and followed the project's progress, taking notice of its numerous daily conflicts and continuous technical challenges and problems. By the end of 2016, the project was far behind schedule, with only 15 kilometres of road completed. Furthermore, the works were to be suspended due to defects in the road layout and exhaustion of the project's budget.

The everyday obstacles involved in building a road across a steep forest, together with the countless conflicts this project has involved, could be the subject of another chapter. What is of interest here, however, and what I would like to emphasize, is the immutable nature of Reyes' dream, as reflected by its re-enactment a century later by President Uribe. This immutability, I suggest, lay not so much in the sort of projects envisaged by the statesmen and explorer, as in the particular vision and rhetoric in which this dream is rooted. In Uribe's speech, we find no mention of cannibal Indians such as those that 'devoured' Reyes' younger brother; neither are the Amazon forests portrayed as a *terra incognita* untouched by civilization; nor did the speaker captivate the audience with his own delusions of conquest and grandeur. Yet the road, identical to Reyes' ambitious transport plan, is assumed by Uribe as a 'colossal project' aimed to transform a 'backward' and 'isolated' space into a modern infrastructural landscape of interoceanic highways and waterways, massive container ports and development hubs. As in Reyes' Amazon, the region's backwardness strikingly contrasts with the inexhaustible resources it encloses; and, although the indigenous peoples are not deemed cannibals anymore, they, along with guerrillas and other outlaw characters 'hiding' in the forests, are still seen as an obstacle to progress and development, as is the topography of the Andean-Amazon *piedemonte*. Against this backdrop, the project itself, far from being a caprice of the megalomaniac explorer or the statesman 'obsessed with progress' – as Uribe himself depicted Reyes – appears to embody the civilizing mission of the state.

We could cite many other examples of this rhetoric in government, development, media and even academic discourses and representations of the territories lying outside the 'control of the state', often pictured as *no man's lands* inhabited by terrorists, drug traffickers and a wide variety of lawless characters. An analysis of the temporal and spatial mutations of discourse in which these territories and its inhabitants are embedded has been beyond the scope of this work. What I have sought to illustrate through the history and ethnography of the road is how this rhetorical construction, which has historically opposed the idea of 'civilization' to 'savagery', 'progress' to 'backwardness' or 'centre' to 'periphery', directly alludes to one of the central axioms in which the project of the state has been founded and sustained.

The image of the Amazon as a frontier space is inextricably bounded to those dichotomous constructions. As noted, this image was constructed and has largely been governed by two dominant tropes. The first sees this region as a vast territory abounding in unexploited resources and vacant lands. This trope is as strong in Codazzi's mid-nineteenth century promising depictions of the *Territorio del Caquetá*, as in present-day allusions to this same territory as 'the last frontier', used both in reference to its untapped resources (biodiversity, oil, mining, tourism) and the dangers (environmental, social, ethnic) involved in present and future plans to exploit them. The other trope, equally enduring, describes this same region as a 'marginal territory' and is conveyed through its various associated meanings: chaos, illegality, isolation, neglect, barbarism, savagery. This trope, furthermore, is habitually expressed in terms of exclusion from the state, or the idea of the frontier as those territories where the state has not arrived or is yet to arrive because it has historically been too weak or unwilling to do so. The main effect of this 'state absence' is that the association of the frontier with lawlessness and barbarity is reinforced or rendered natural. The frontier, then, appears as a 'natural' theatre of conflict where the 'vacuum' of the state is filled with and disputed between guerrillas, drug dealers, and all sorts of outlaw characters.

These two tropes are fundamental to understanding past and contemporary state practices and policies in the numerous 'peripheries', 'margins' and 'frontiers' of the nation. They manifest themselves in the many projects, real and utopian, aimed at 'civilizing' or 'developing' these territories, as well as in the quotidian images of a development that appears elusive, the same images through which frontier peoples condemn the absence of the state or convey their memories and feelings of marginality, abandonment or exclusion from it. I have focused on the subject of roads because few infrastructures so powerfully embody these feelings and memories. Consider, for example, Guillermo's analogy between the *trampoline of death*'s hairpin curves and the 'twisted and treacherous' nature of Colombians (Chapter 4); or the enduring image of the road as a state plot to imprison the inhabitants of the Putumayo and the *colonos*' countless complaints to the state for neglecting 'weak peoples' (Chapter 3); or the distrust towards the government from peasant communities inhabiting the San Francisco-Mocoa road project area (Chapter 5). These images, claims and feelings reflect real forms of spatial, social and political exclusion. Yet, if we situate these forms of exclusion in the long-term or think of them in terms of their permanence in time and embeddedness in space, as I have attempted to do here, they reappear in a different guise, that is, not as manifestations of exclusion *from* but as expressions of inclusion *in* the state.

In emphasizing this distinction between exclusion and inclusion, I have sought to show that the Colombian Amazon does not constitute a territory lying outside the state. Rather, what this distinction means is that the incorporation of this territory to the political and spatial order of the state has depended on the perpetuation of its *status* of frontier or its image as a space existing in a relationship of exteriority and opposition to this order. Following Agamben's notion of exception, I have employed the term 'inclusive exclusion' to describe this relationship and the different ways in which it has shaped the spatial history of the frontier. Throughout the book, I drew attention to several instances, characters, and events where this relationship manifests itself: the nineteenth-century *criollos*' narratives and cartographical constructions of the country's territory; Reyes' chaotic map of cannibals, minerals, rubber and grand infrastructural projects; the Capuchin road's solemn inauguration rituals upon breaking the Andean 'barrier' separating 'civilization' from 'savagery'; the Sibundoy Valley's cadastral plans exhibiting the rampant dispossession of indigenous lands; the language that frames conflicts and disputes involved in the construction of the Pasto-Puerto Asís road; the *trampoline of death*'s endless record of deadly 'accidents'; and the representations of the 'peasant' and the 'displaced' in the context of the road project, just to name a few.

Although these instances and events are confined to a geographically delimited space, they also reflect the highly heterogeneous and uneven character of this space. Thus, I have suggested that in reflecting upon the notion of frontiers, we should think of a concept whose meaning transcends or is not restricted to a fixed territory or geographical area, and can be better defined as a *condition* embodying a relationship of inclusive-exclusion, regardless of the ways in which this relation is encountered or expressed. Beyond the context of the road, there are many other everyday situations in Colombia where this condition of frontier is evident. For instance, the sporadic practice, often carried out by authorities themselves, of 'exporting' or sending homeless or 'marginal' persons from major cities to peripheral regions; the ID checks in bus terminals and airports in peripheral regions (often considered 'conflict zones') aimed at searching for people's criminal records, a practice that brings to mind and reinforces the old image of such regions as prisons or places of exile; or the government's immemorial habit of stigmatizing the inhabitants of these same regions as subversive subjects.

Along with these practices, there are others no less pervasive yet more brutal where this condition is encountered. I would like to bring up one of them, for I think it fully discloses the violence that characterizes this condition. It refers to the so-called 'false positives', a practice involving the Colombian military which consisted of murdering civilians and presenting them as guerrillas killed in combat. This practice became a scandal of international proportions in 2008, during the second term of Alvaro Uribe's administration. The scandal exploded at the end of that year, when 19 young men from peripheral areas of Bogotá, whose families had reported as missing, were found dead and buried as 'NN' (unknown) about 700 kilometres away, in the northern department of Santander (Revista Semana 2008). The youths, as it was later established, had been tricked by a 'recruiter' who offered them jobs and then delivered them to army forces who subsequently murdered and falsely reported them as casualties. The public outcry that resulted from this case was followed by many similar claims across the country that revealed the prevalence of this practice.

Although the different reports that followed the 2008 scandal noted that this type of extrajudicial executions was not new, they emphasized the massive increase of cases during Uribe's two presidential terms (e.g. CCEEU 2012; CINEP 2011; FIDH 2012; UNHRC 2010). For instance, according to figures from the Coordination Colombia-Europe-United States Human Rights Observatory (CCEEU), between 1994 and 2001 the number of extrajudicial executions by the state's security forces amounted to 739, while in the period from 2002 to 2010 that number increased to 3,512 (CCEEU 2012, p.8). This rise has been widely associated with Uribe's counter-insurgency campaign, which put pressure on the military to show results and also devised a system that gave rewards and incentives to army units and commanders producing more casualties. The systematic character of the killings was revealed and often stressed in reports in relation to the profiles of the victims and the *modus operandi* of the killings. The former were mostly young men that would easily fit the profile of combatants, generally from low-income families living in rural areas or zones affected by the conflict, although indigents, homeless, displaced, disabled and other individuals regarded as 'marginal' or 'lawless' were also common 'targets'.

The *modus operandi*, as summarized in one of the reports (FIDH 2012, pp.19–20), was the following: in some cases, the victims were detained arbitrarily and without judicial warrant in the place where they lived or worked, whilst in others they were previously selected by a paid 'recruiter' or 'informant' who tricked them with a job offer or another false promise. Once detained or recruited, the victim or victims were transported (often long distances so they could not be identified by local witnesses) and delivered to a military unit who executed them; in most cases, the executions took place in remote rural areas where counter-insurgency operations were taking place. Following the execution, the crime scene was manipulated to simulate a combat: the victim was stripped of his clothes and IDs, dressed in military uniform and fitted with weapons; subsequently, pictures of the scene were taken by the same executioners as evidence of the 'combat' (pictures which often revealed the crude simulation of the crime scene in details such as victims wearing boots on the wrong feet or clothes that were too big). The victim was then buried as 'NN' and then reported as a casualty to the respective military command or battalion. Finally, the commander and soldiers of the military unit responsible for the casualty or casualties were rewarded with leave, economic compensations or congratulations.

Allow me to quote an example of a 'false positive' included in a report that provides a chronological sample of 951 cases of individual and collective killings built on complaints between 1984 and 2011. The case, listed as Case no. 0819, took place on 20th January 2008 in Puerto Asís (Putumayo) and reads as follows:

> In Puerto Asís, Putumayo, troops from the Mobile Brigade 13 executed the peasant leaders Hugo Armando Torres and Heynar Alexander Guerrero Paredes, in events which took place in the *vereda* Nueva Unión, settlement of Teteyé. According to the complaint: 'the peasants were executed in an extrajudicial way, when members of the Brigade XIII burst into the home of Hugo Armando Torres and detained him arbitrarily. The soldiers asked the inhabitants for the leader and despite them identifying him as a leader from the community, one of the soldiers accused him of being a "sapo guerrillero" [snitch]. Minutes later, the military took Hugo Armando to the outskirts of the settlement and half-an-hour later, approximately, rifle shots and two explosions were heard. Subsequently, a military helicopter circled overhead and fired indiscriminately over the families inhabiting the place. Hours later, two more helicopters flew over. One of them landed near the house of Heynar Alexander Guerrero, his house was broken into, and since that moment no news was heard of the young man. On the following day the lifeless bodies of Hugo Armando Torres and Heynar Alexander Guerrero were found in the Puerto Asís morgue. The corpses, showing signs of torture, were buried in the cemetery as NN in the same crypt and without a coffin. The death of the two community leaders was reported as the result of an armed combat with guerrilleros from the Front 48 of the FARC' (CINEP 2011, pp.180–181).

The scandal and intense debate unleashed by the 'false positives' can be explained by the fact that they constituted a form of violence against civilians that, different to other forms of conflict-related violence, revealed the direct involvement and responsibility of the state's military forces. The way in which this involvement has been usually denounced by human rights and media reports on the subject, is by placing heavy emphasis on the systematic and widespread character of the executions, together with the impunity and cover-up surrounding the majority of such crimes. This emphasis has been critical in making visible the magnitude of the phenomenon, and in gaining the attention of institutions like the International Criminal Court. Much less emphasis has been placed, however, on the question of what made the victims of 'false positives' expendable or, alternatively, what rendered these extrajudicial executions such a natural or spontaneous response to the pressure on the military to 'show results'. This is a question that is seldom asked because it seems redundant or self-evident, because in the country's long-history of conflict, the status of potential victim or 'target' of 'false positives', forced displacement, disappearance and so forth has, in most cases, been defined not by the person's involvement in the conflict, but by his pre-existent (or assumed) condition of 'marginal', 'excluded', 'vulnerable' or any other similar adjective. Put differently, this condition – to quote Agamben's statement on the character of sovereign violence – is what allows killing *without committing homicide and without celebrating a sacrifice*' (Agamben 1998, p.83, emphasis in original).

I bring up the 'false positives' story here partly because it speaks about the relationship between violence and exception so ingrained in Colombia's conflict. The genesis of this conflict can only be fully grasped if we look at how this relationship has been built and maintained through the production of physical and symbolic margins and frontiers within the body politic of the nation state. The 'false positives' are in this sense nothing but one of several manifestations of the symbolic and physical violence through which certain 'excluded' territories and populations are incorporated into the order of the state. But the 'false positives' are also, in a wider sense, an expression of a typology of violence deeply rooted in long-established global bordering practices suggesting that the state of exception has become a dominant form of government (Agamben 2005). Well-known and researched contemporary examples falling within this typology include the counter-terrorism 'targeting' practices involved in the US drone campaigns, which often result in the incidental killing of civilians and other kinds of 'collateral damage'; the 'coercive interrogation' techniques in extraterritorial detention facilities; or the securing and confinement policies in refugee and asylum seekers' camps.

As various scholars have noted (e.g. Gregory 2006, 2014; Minca and Vaughan-Williams 2012; Shaw and Akhter 2012) these bordering practices are enabled by and strongly dependent on the creation of different sorts of exceptional laws and liminal spaces: buffer zones, *no man's lands*, frontiers, and so forth. However, as noted at the beginning of the book, what ultimately renders such practices as a structural element within the *modus operandi* of states is the fact that they reflect entrenched socio-spatial orderings built along geopolitical, racial, class and gender lines. The hierarchical structuring of these categories is expressed in a myriad of intangible or symbolic barriers that lie at the core of the state project, and underpin the different technologies and infrastructures through which this project is crafted.

In discussing the genesis of state power, Pierre Bourdieu states that 'the most brutal relations of force are always simultaneously symbolic relations' (Bourdieu 1994, p.12). The reason for this, Bourdieu argues, is that symbolic relations are extremely difficult to break or alter since they are rooted in cognitive structures, structures through which the social world is constructed. The state, moreover, plays a central role in shaping such structures for it 'establishes and inculcates common forms and categories of perception and appreciation, social frameworks of perception, of understanding or of memory, in short *state forms of classification*' (Bourdieu 1994, p.13, emphasis in original). The condition of the frontier, of being included by means of exclusion, is firmly grounded in these relations. To account for the everyday actions and struggles through which this condition is contested is fundamental in order to apprehend the world as it is lived rather than merely thought or discursively constructed. Yet, an equally significant aspect of this frontier condition, and what makes it brutal is, I think, its permanence in time and embeddedness in space. This is so because through this permanence and embeddedness the violence this condition embodies becomes normalized, even to the point that we come to see it as immanent or natural. To confront this violence, then, it is not enough to condemn its visible or material manifestations and effects. This task also, and necessarily, requires us to critically investigate and question the symbolic boundaries that sustain it and through which it is perpetuated.

References

Abrams, Philip. 2006. "Notes on the difficulty of studying the state", in *The Anthropology of the State: A reader*, edited by Aradhana Sharma and Akhil Gupta, 112–130. Oxford: Blackwell.

Agamben, Giorgio. 1998. *Homo Sacer. Sovereign power and bare life*. Stanford: Stanford University Press.

Agamben, Giorgio. 2005. *State of exception*. Chicago: The University of Chicago Press.

Aiton, Arthur. 1994 [1940]. "Latin-American frontiers", in *Where Cultures Meet: Frontiers in Latin American history*, edited by David Weber and Jane Rausch, 19–25. Wilmington, DE: Jaguar Books.

Allen, John. 2011. "Topological twists: Power's shifting geographies", *Dialogues in Human Geography*, Vol.1 (3): 283–298.

Anderson, Benedict. 2006. *Imagined Communities. Reflections on the origins and spread of nationalism*. London & New York: Verso.

André, Edouard. 1877. "L'Amérique équinoxiale", *Le tour du Monde*, Vol.34: 273–368.

André, Edouard. 1984. *América equinoccial, Colección América Pintoresca*, vol.2. Cali: Carvajal.

Aparicio, Juan Ricardo. 2005. "Intervenciones etnográficas a propósito del sujeto desplazado: estrategias para (des)movilizar una política de la representación", *Revista Colombiana de Antropología*, Vol.41: 135–169.

Arendt, Hannah. 1951. *The Origins of Totalitarianism*. New York: Harcourt, Brace and Company.

Arias, Julio. 2005. *Nación y diferencia en el siglo XIX colombiano: orden nacional, racialismo y taxonomías poblacionales*. Bogotá: Uniandes-Ceso.

Frontier Road: Power, History, and the Everyday State in the Colombian Amazon, First Edition. Simón Uribe.
© 2017 John Wiley & Sons Ltd. Published 2017 by John Wiley & Sons Ltd.

Ariza, Eduardo, María Clemencia Ramírez and Leonardo Vega. 1998. *Atlas Cultural de la Amazonia Colombiana. La Construcción del territorio en el siglo XX*. Bogotá: ICAN.

Banco Agrario de Colombia. n.d. "Gerencia de vivienda. Resumen General del Proyecto. Programa retorno o reubicación desplazados. Convocatoria 200508. Radicación 1743054421". Bogotá: Banco Agrario de Colombia.

Barnhart, Donald. 1985. "Colombian transport and the reforms of 1931: An evaluation", *Hispanic American Historical Review*, Vol.38: 1–24.

Belaúnde, Victor. 1994. "The frontier in Hispanic America", in *Where Cultures Meet. Frontiers in Latin American history*, edited by David Weber and Jane Rausch, 33–41. Wilmington, DE: Jaguar Books.

Belcher, Oliver et al. 2008. "Everywhere and nowhere: The exception and the topological challenge to geography," *Antipode*, Vol.40 (4): 499–503.

Bell, Morag, Robin Butlin and Michael Heffernan, eds. 1995. *Geography and Imperialism. 1820–1940*. Manchester: Manchester University Press.

Benjamin, Walter. 1969. *Illuminations: Essays and reflections*, edited by Hannah Arendt, New York: Schocken.

Bergquist, Charles. 1986. *Coffee and Conflict in Colombia, 1886–1920*. Durham, NC: Duke University Press.

Boletín de la Provincia de Nuestra Señora de la Candelaria de Colombia. 1924. Vol.II, no.XXVII, Bogotá: Imprenta San Bernardo.

Bolívar, Ingrid. 1999. "Sociedad y estado: la configuración del monopolio de la violencia", *Revista Controversia*, 175: 9–39.

Bolívar, Ingrid. 2003. *Violencia política y formación del estado. Ensayo historiográfico sobre la dinámica de la Violencia de los cincuenta en Colombia*. Bogotá: CINEP, Ceso, Uniandes.

Bonilla, Víctor. 1972. *Servants of God or Masters of Men? The story of a Capuchin mission in Amazonia*. London: Penguin.

Bonilla, Víctor. 2006. *Siervos de Dios y Amos de los Indios. El Estado y la Misión Capuchina en el Putumayo*. Popayán: Editorial Universidad del Cauca.

Bourdieu, Pierre. 1994. "Rethinking the state: Genesis and structure of the bureaucratic field", *Sociological Theory*, Vol.12 (1): 1–18.

Braudel, Fernand. 1960. "History and the social sciences: The long duration", *Political Research, Organization and Design*, Vol.3 (5): 3–13.

Brücher, Wolfgang. 1968. *La colonización de la selva pluvial en el piedemonte amazónico de Colombia. El territorio comprendido entre el Ariari y el Ecuador*. Bogotá: IGAC.

Brücher, Wolfgang. 1970. "La colonización de la selva pluvial en el piedemonte amazónico de Colombia", Instituto Colombo Alemán, *Investigación Científica*, 4: 97–123.

Bushnell, David. 1993. *The Making of modern Colombia. A nation in spite of itself*. Berkeley: University of California Press.

Cajiao, Leopoldo. 1900. *Arrendamiento o venta del Territorio del Caquetá*. Bogotá: Imprenta de la Luz.

Caldas, Francisco José. 1966[1808]. *Obras completas de Francisco José de Caldas*. Bogotá: Universidad Nacional de Colombia, Imprenta Nacional.

Calderón, Florentino. 1902. *Nuestros desiertos del Caquetá y del Amazonas y sus riquezas*. Bogotá: Imprenta de Luis M. Holguín.

Campbell, Jeremy. 2012. "Between the material and figural road: The incompleteness of colonial geographies in Amazonia", *Mobilities*, Vol.7 (4): 481–500.

Campbell, Jeremy and Kregg Hetherington. 2014. "Nature, infrastructure, and the state: Rethinking development in Latin America", *The Journal of Latin American and Caribbean Anthropology*, Vol.19 (2): 191–194.

Carroll, Patrick. 2006. *Science, Culture, and Modern State Formation*. Berkeley: University of California Press.

Casas, Justo. 1999. *Evangelio y Colonización. Una aproximación a la historia del Putumayo desde la época prehispánica hasta el presente*. Bogotá: Ecoe.

CCEEU. 2012. "Ejecuciones extrajudiciales en Colombia 2002–2010. Crímenes de lesa humanidad bajo el mandato de la política de defensa y seguridad democrática", *Documentos temáticos*, 12. Bogotá: Códice.

Chaparro, Ana María. 2015. *Visión de futuro indígena y su incidencia en el desarrollo en Colombia. El caso de las comunidades inga y kamëntza (Putumayo) ante la Iniciativa para la Integración de la Infraestructura Regional Suramericana (IIRSA)*. Bogotá: Cider-Universidad de los Andes.

Chatterjee, Partha. 2004. *The Politics of the Governed. Reflections on popular politics in most of the world*. New York: Columbia University Press.

Chaves, Margarita. 1998. "Identidad y representación entre indígenas y colonos de la Amazonia occidental Colombiana", in *Identidad, modernidad y desarrollo*, edited by María Lucía Sotomayor, 283–296. Bogotá: Ican-Colciencias.

Chaves, Milciades. 1945. "La colonización de la Comisaría del Putumayo. Un problema etno-económico-geográfico de importancia nacional", *Boletín de Arqueología*, ICAN, Vol.1 (1): 567–598.

CINEP. 2011. *Deuda con la humanidad II. 23 años de falsos positivos (1988–2011)*, CINEP-PPP. Bogotá: Códice.

Ciro, Alejandra. 2009. "De la selva a la pradera: reconfiguración espacial del piedemonte caqueteño 1950–1965", Documento CESO no. 159. Bogotá: Universidad de los Andes.

Codazzi, Agustín. 1996. *Geografía física y política de la Confederación Granadina. Estado del Cauca, Territorio del Caquetá. Obra dirigida por el General Agustín Codazzi*, edited by Camilo Domínguez, Augusto Gómez and Guido Barona. Bogotá: COAMA, FEN, IGAC.

Codazzi, Agustín, Manuel Paz and Felipe Pérez. 1889. "Carta de la Nueva Granada dividida en provincias, 1832 a 1856. Uti possidetis de 1810", *Atlas geográfico e histórico de la República de Colombia*. Paris: A. Lahure.

CODHES. 2014. "El desplazamiento forzado y la imperiosa necesidad de la paz". Available from: http://www.codhes.org/index.php?option=com_si&type=4 [Accessed 7th October 2016].

Colombijn, Freek. 2002, "Introduction: On the road", *Bijdragen tot de Taal-, Land-en Vokenkunde*, Vol.158 (4): 595–617.

Coronil, Fernando. 1997. *The Magical State. Nature, money, and modernity in Venezuela*. Chicago & London: The University of Chicago Press.

Corte Constitucional. 2004. Sentencia T025/04. Available from http://www.corteconstitucional.gov.co/relatoria/2004/t-025-04.htm [Accessed 7th October 2016].

Corte Constitucional. 2009. Auto 008/09. Available from: http://www.corteconstitucional.gov.co/RELATORIA/Autos/2009/A008-09.htm [Accessed 7th October 2016].

Crist, Raymond and Ernesto Guhl. 1947. "Pioneer settlement in eastern Colombia", *Annual Report of the Board of Regents of the Smithsonian Institution, Publication 4272*, 391–414. Washington, DC: United States Government Printing Office.

Crist, Raymond and Charles Nissly. 1973. *East from the Andes*. Pioneer settlement in the South American heartland. Gainesville: University of Florida Press.

Dalakoglou, Dimitris. 2012. "'The road from capitalism to capitalism': Infrastructures of (post) socialism in Albania", *Mobilities*, Vol.7 (4): 571–586.

Das, Veena and Deborah Poole. 2004. "State and its margins: Comparative ethnographies", in *Anthropology in the Margins of the State*, edited by Veena Das and Deborah Poole, 3–34. Santa Fe: School of American Research Press.

De Certeau, Michel. 1998. *The Practice of Everyday Life*. Los Angeles: University of California Press.

De Pinell, Gaspar and Canet del Mar. 1924. *Relaciones interesantes y datos históricos sobre las misiones católicas del Caquetá y Putumayo desde el año 1632 hasta el presente*. Bogotá: Imprenta Nacional.

Domínguez, Camilo. 1984. "National expansion and development policies in the Colombian Amazon", in *Frontier Expansion in Amazonia*, edited by Marianne Schmink and Charles H. Hood, 405–418. Gainesville: University of Florida Press.

Domínguez, Camilo. 2005. *Amazonia Colombiana: Economía y poblamiento*. Bogotá: Universidad Externado de Colombia.

Driver, Felix. 2001. *Geography Militant. Cultures of exploration and empire*. Oxford: Blackwell.

Duncan, Silvio and John Markoff. 1978. "Civilization and barbarism: Cattle frontiers in Latin America", *Comparative Studies in Society and History*, Vol.20 (4): 587–620.

Durán, David et al. 2007. *Desplazamiento forzado en Colombia. Derechos, acceso a la justicia y reparaciones*, ACNUR, CEDHUL, Generalitat Valenciana. Colombia: Futura Impresores.

Easterday, Jennifer. 2008. "Litigation or legislation: Protecting the rights of internally displaced persons in Colombia". Unpublished paper. Available from: http://works.bepress.com/jennifer_easterday/1 [Accessed 7th October 2016].

El Tiempo. 1991. "Tragedia en Putumayo: las víctimas podrían pasar de 50". 21st July. Available from: http://www.eltiempo.com/archivo/documento/MAM-123694 [Accessed 7th October 2016].

El Tiempo. 1996. "Vía Pasto-Mocoa: 148 Km De Miedo". 3rd November. Available from: http://www.eltiempo.com/archivo/documento/MAM-573009 [Accessed 7th October 2016].

Fairhead, James. 1992. "Paths of Authority: Roads, the state and the market in eastern Zaire", *The European Journal of Development Research*, Vol.4 (2): 17–35.

Fajardo, Darío. 1996. "Fronteras, colonizaciones, y construcción social del espacio", in *Frontera y poblamiento: Estudios de historia y antropología de Colombia y Ecuador*, edited by Chantal Caillavet and Ximena Pachón, 237–282. Bogotá: IFEA, Instituto Amazónico de Investigaciones Científicas, Universidad de los Andes.

Fearnside, Philip. 2007. "Brazil's Cuiabá-Santarem (BR-163) highway: The environmental cost of paving a soybean corridor through the Amazon", *Environmental Management*, Vol.39 (5): 601–614.

FIDH. 2012. *Colombia. The war is measured in litres of blood. False positives, crimes against humanity: Those most responsible enjoy impunity*. Available from: http://www.fidh.org/IMG/pdf/rapp_colombie__juin_2012_anglais_def. pdf [Accessed 7th October 2016].

Flórez, Margarita. 2007. *Selva Abierta: Vía Pasto-Mocoa e hidrovía del Putumayo: expresiones en Colombia de la Iniciativa para la Integración de la Infraestructura Regional Suramericana*. Bogotá: BIC.

Forero, Luis. 1928. *La Pedrera: relato del combate entre colombianos y peruanos en el año de 1911*. Bogotá: Editorial Bolívar.

Foucault, Michel. 1991. "Politics and the study of discourse", in *The Foucault Effect. Studies in governmentality*, edited by Graham Burchell, Colin Gordona and Peter Miller, 53–72. Chicago: The University of Chicago Press.

Foucault, Michel. 2007. *Security, Territory, Population. Lectures at the College de France 1977–1978*. London: Palgrave.

Foweraker, Joe. 1981. *The Struggle for Land: A political economy of the pioneer frontier in Brazil, 1930 to the present*. Cambridge: Cambridge University Press.

Friede, Juan. 1945. "Leyendas de Nuestro Señor de Sibundoy y el Santo Carlos Tamabioy", *Boletín de Arqueología*, Vol.1 (4): 315–318.

Gallup, John Luke, Alejandro Gaviria and Eduardo Lora. 2003. *Is Geography Destiny? Lessons from Latin America*. Stanford, CA: Inter-American Development Bank, Stanford University Press and World Bank.

García, Clara. 2003. "Enfoques y problemas de la investigación sobre territorios de frontera interna en Colombia", in *Fronteras. Territorios y Metáforas*, edited by Clara García, 47–60. Medellín: INER, Universidad de Antioquia, Hombre Nuevo Editores.

García, Mauricio and José Rafael Espinosa, 2011. "Estado, municipio y geografía", in Mauricio García et al., *Los Estados del país. Instituciones municipales y realidades locales*, 52–105. Bogotá: Centro de Estudios de Derecho, Justicia y Sociedad, Dejusticia.

García, Mauricio et al. 2011. *Los Estados del país. Instituciones municipales y realidades locales*. Bogotá: Centro de Estudios de Derecho, Justicia y Sociedad, Dejusticia.

Godlewska, Anne and Neil Smith, eds. 1994. *Geography and Empire*. Oxford: Blackwell.

Gómez, Augusto. 1993. "Traición a la Patria", *Revista Universitas*, 37: 6–24.

Gómez, Augusto. 1996. "Bienes, rutas y mercados (siglos XV–XIX). Las relaciones de intercambio entre las tierras bajas de la Amazonia y las tierras altas de los Andes", *Revista de Antropología y Arqueología*, Vol.IX (1–2): 51–80.

Gómez, Augusto. 2005. "El Valle de Sibundoy: el despojo de una heredad. Los dispositivos ideológicos, disciplinarios y morales de dominación", *Anuario colombiano de historia social y de la cultura*, 32: 51–73.

Gómez, Augusto. 2011. *Putumayo. Indios, misión, colonos y conflictos (1845–1970)*. Popayán: Editorial Universidad del Cauca.

Gómez, Augusto and Camilo Domínguez. 1995. "Quinerías y caucherías de la Amazonia. Caminos y varaderos de la Amazonia", in *Caminos Reales de Colombia*, edited by Mariano Useche. Bogotá: Fondo FEN.

Gómez, Augusto, Ana Lesmes and Claudia Rocha. 1995. *Caucherías y conflicto Colombo-Peruano. Testimonios 1904–1934*. Bogotá: Disloque Editores.

González, Fernán. 1977. "Consolidación del Estado Nacional", *Controversia*, 59–60: 1–148.

González, Fernán, Ingrid Bolívar and Teófilo Vásquez. 2003. *Violencia Política en Colombia. De la nación fragmentada a la construcción del Estado*. Bogotá: CINEP.

Gramsci, Antonio. 1971. *Selections from the Prison Notebooks*, edited and translated by Quintin Hoare and Geoffrey Nowell Smith. New York: International Publishers.

Gregory, Derek. 2006. "The Black Flag: Guantánamo Bay and the space of exception", *Geografiska Annaler: Series B*, Vol.88 (4): 405–427.

Gregory, Derek. 2014. "Drone geographies", *Radical Philosophy*, Vol.183: 7–19.

Guhl, Ernesto. 1976. *Colombia: bosquejo de su geografía tropical*, Vol.2. Bogotá: Instituto Colombiano de la Cultura.

Guhl, Ernesto. 1991. *Escritos geográficos. Las fronteras políticas y los límites naturales*. Bogotá: Fondo FEN.

Gutiérrez, Rufino. 1921. *Monografías de Rufino Gutiérrez. Caquetá y Putumayo (Informes oficiales del procurador de hacienda. Caquetá y Putumayo, 1912)*, Vol.I. Bogotá: Imprenta Nacional.

Hansen, Thomas and Finn Stepputat, eds. 2001. *States of Imagination. Ethnographic explorations of the postcolonial state*. Durham, NC: Duke University Press.

Hansen, Thomas and Finn Stepputat. 2005. "Introduction", in *Sovereign Bodies. Citizens, migrants, and states in the postcolonial world*, edited by Thomas Blom Hansen and Finn Stepputat, 1–36. Princeton & Oxford: Princeton University Press.

Harley, John Brian. 2001. *The New Nature of Maps. Essays in the history of cartography*. Baltimore, MD: John Hopkins University Press.

Harvey, Penny. 2001. "Landscape and commerce: Creating contexts for the exercise of power", in *Contested Landscapes. Movement, exile and place*, edited by Barbara Bender and Margot Winer, 197–210. Oxford & New York: Berg.

Harvey, Penny. 2005. "The materiality of state-effects: An ethnography of a road in the Peruvian Andes", in *State Formation. Anthropological perspectives*, edited by Christian Krohn-Hansen and Knut Nustad, 123–141. London: Pluto Press.

Harvey, Penny. 2012. "The topological quality of infrastructural relation: An ethnographic approach", *Theory, Culture & Society*, Vol.29 (4–5): 76–92.

Harvey, Penny. 2014. "Infrastructures of the frontier in Latin America", *The Journal of Latin American and Caribbean Anthropology*, Vol.19 (2): 280–283.

Harvey, Penny and Hannah Knox. 2008. "'Otherwise engaged'. Culture, deviance and the quest for connectivity through road construction", *Journal of Cultural Economy*, Vol.1 (1): 79–92.

Harvey, Penny and Hannah Knox. 2012. "The enchantments of infrastructure", *Mobilities*, Vol.7 (4): 521–536.

Harvey, Penny and Hannah Knox. 2015. *Roads: An anthropology of infrastructure and expertise*. New York: Cornell University Press.

Hegel, Wilhelm Friedrich. 1976[1821]. *Philosophy of Right*. London: Oxford University Press.

Hegen, Edmund. 1963. "The Andean cultural frontier", *Journal of Inter-American Studies*, Vol.5 (4): 431–436.

Hegen, Edmund. 1966. Highways into the Upper Amazon Basin. *Pioneer lands in Southern Colombia, Ecuador, and Northern Peru*. Gainesville: University of Florida Press.

Hobbes, Thomas. 1937[1651]. *Leviathan*. London: J.M. Dent & Sons Ltd.

Horna, Hernán. 1982. "Transportation, modernization and entrepreneurship in nineteenth century Colombia", *Journal of Latin American Studies*, Vol.14 (1): 33–53.

IADB. 2009a. "San Francisco-Mocoa alternate road construction project-Phase I. Loan Proposal". Available from: http://idbdocs.iadb.org/wsdocs/getdocument.aspx?docnum=35025319 [Accessed 7th October 2016].

IADB. 2009b. "Informe de Gestión Ambiental y Social (IGAS). Corredor vial Pasto-Mocoa. Variante San Francisco-Mocoa (CO-L1019)". Available from http://idbdocs.iadb.org/wsdocs/getdocument.aspx?docnum=2222785 [Accessed 7th October 2016].

IADB. 2009c. "Análisis, identificación y propuesta de instrumentos legales de ocupación del suelo en el área de influencia de la variante San Francisco-Mocoa", in *Informe de Gestión Ambiental y Social (IGAS). Corredor vial Pasto-Mocoa. Variante San Francisco-Mocoa (CO-L1019)*". Annex document. Available from: http://idbdocs.iadb.org/wsdocs/getdocument.aspx?docnum=2222785 [Accessed 7th October 2016].

IIRSA. 2011. Project Portfolio 2011. XVIII Meeting of National Coordinators, Rio de Janeiro, Brazil. Available from: http://www.iirsa.org/admin_iirsa_web/Uploads/Documents/cnr18_rio11_presentacion_cartera_eng.pdf [Accessed 7th October 2016].

INCOPLAN. 2008. "Elaboración del Plan Básico de Manejo Ambiental y Social (PBMAS) de la Reserva Forestal Protectora de la Cuenca Alta del Río Mocoa". 5 vols. BID, Corpoamazonia, Invías. Available from http://www.iadb.org/es/proyectos/project-information-page,1303.html?id=co-l1019#doc [Accessed 15th November 2011].

Informe sobre las Misiones del Putumayo. 1916. Bogotá: Imprenta Nacional.

Informes sobre las Misiones del Caquetá, Putumayo, Guajira, Casanare, Meta, Vichada, Vaupés y Arauca. 1917. Bogotá: Imprenta Nacional.

Iza, Delfín. 1924. "'Amazonia Colombiana. El peligro existe' and 'Asuntos del Caquetá y Putumayo. Es ignorancia de la geografía o abandono de nuestros intereses'", *Asuntos de actualidad relativos al Putumayo*. Bogotá: Casa Editorial Marconi.

James, Preston E. 1923. "The transportation problem of highland Colombia", *Journal of Geography*, Vol.22 (9): 346–357.

James, Preston E. 1941. "Expanding frontiers of settlement in Latin America", *Hispanic American Historical Review*, Vol.21 (2): 183–195.

Jaramillo, Jaime. 1984. "Nación y región en los orígenes del Estado nacional en Colombia", in *Problemas de la formación del estado y de la nación en Hispanoamérica*, edited by Inge Buisson et al. Bonn: Inter Nationes.

Jessop, Bob. 2009. *State Power. A strategic-relational approach*. Cambridge, UK: Polity Press.

Kernaghan, Richard. 2009. *Coca's Gone. Of might and right in the Huallaga post-boom*. Stanford, CA: Stanford University Press.

Kernaghan, Richard. 2012. "Furrows and walls, or the legal topography of a frontier road in Peru", *Mobilities*, Vol.7 (4): 501–520.

Kirskey, Eben and Kiki van Bilsen, 2002. "A road to freedom. Mee articulations and the Trans-Papua Highway", *Bijdragen tot de Taal-, Land-en Vokenkunde*, Vol.158 (4): 837–854.

König, Hans-Joachim. 1984. "Símbolos nacionales y retórica política en la Independencia: el caso de la Nueva Granada", in Inge Buisson et al., *Problemas de la formación del Estado y de la nación en Hispanoamérica*. Bonn: Inter Nationes.

Krohn-Hansen, Christian and Knut Nustad, eds. 2005. *State Formation. Anthropological perspectives*. London: Pluto Press.

Labor de los Misioneros en el Caquetá y Putumayo, Magdalena y Arauca. Informes años 1918–1919. 1919. Bogotá: Imprenta Nacional.

Larson, Brooke. 2004. *Trials of Nation Making. Liberalism, race, and ethnicity in the Andes, 1810–1910*. Cambridge: Cambridge University Press.

Las Misiones en Colombia. Obra de los misioneros Capuchinos en el Caquetá y Putumayo. 1912. Bogotá: Imprenta de la Cruzada.

Law 387 of 1997. Available from: https://www.brookings.edu/wp-content/uploads/2016/07/Colombia_Law387_1997_Eng.pdf [Accessed 10th October 2016].

Lefebvre, Henri. 1991. *The Production of Space*. Oxford: Blackwell.

Lefebvre, Henri. 2009. "Space and the state", in *Henri Lefebvre. State, space, world: Selected essays*, edited by Neil Brenner and Stuart Elden, 223–253. Minneapolis and London: University of Minnesota Press.

Legg, Stephen and Alex Vasudevan. 2011. "Introduction: Geographies of the Nomos", in *Spatiality, Sovereignty and Carl Schmitt: Geographies of the nomos*, edited by Stephen Legg, 1–23. London: Routledge.

Legrand, Catherine. 1986. *Frontier Expansion and Peasant Protest in Colombia, 1850–1936*. Albuquerque: University of New México Press.

Lemaitre, Eduardo. 1981[1951]. *Rafael Reyes. Biografía de un Gran Colombiano*. Bogotá: Banco de la República.

Llanos, Héctor and Roberto Pineda. 1982. *Etnohistoria del gran Caquetá*. Bogotá: Banco de la República.

Locke, John. 1980[1690]. *The Second Treatise of Civil Government*. Cambridge: Hackett.

Londoño, Jaime E. 2003. "La frontera: un concepto en construcción", in *Fronteras. Territorios y Metáforas*, edited by Clara García, 61–86. Medellín: INER, Universidad de Antioquia, Hombre Nuevo Editores.

Lye, Tuck-Po. 2005. "The road to equality? Landscape transformation and the Batek of Pahang, Malaysia", in *Property and Equality, Vol. 2: Encapsulation, commercialisation, discrimination*, edited by Thomas Widlok and Wolde Gossa Tadesse, 90–103. Oxford: Berghahn.

Marichal, Carlos. 2002. *México y las Conferencias Panamericanas 1889–1938*. Ciudad de México: Secretaría de Relaciones Exteriores.

Martínez, Felipe. 2013. "Héroes de la civilización. La Amazonía como cosmópolis agroexportadora en la obra del General Rafael Reyes", *Anuario colombiano de historia social y de la cultura*, Vol.40 (2): 145–177.

Martínez, Frédéric. 1997. "Apogeo y decadencia del ideal de la inmigración europea en Colombia, siglo XIX", *Boletín Cultural y Bibliográfico*, Biblioteca Luís Ángel Arango, Vol. XXXIV (44): 3–45.

Marx, Karl. 1949[1867]. *Capital: A critical analysis of capitalist production*. London: Allen and Unwin.

Masquelier, Adeline. 2002. "Road mythographies: Space, mobility, and the historical imagination in postcolonial Niger", *American Ethnologist*, Vol.29 (4): 829–856.

Mbembe, Achille. 2003. "Necropolitics", *Public Culture*, Vol.15 (1): 11–40.

McFarlane, Anthony. 1993. *Colombia before Independence: Economy, society and politics under Bourbon rule*. New York and Cambridge: Cambridge University Press.

Melo, Jorge Orlando. 1986. "La evolución económica de Colombia, 1830–1900", *Manual de Historia de Colombia*, Vol.II: 133–207. Bogotá: Instituto Colombiano de Cultura.

Merteens, Donny and Margarita Zambrano. 2010. "Citizenship deferred: The politics of victimhood, land restitution and gender justice in the Colombian (post?) conflict", *The International Journal of Transitional Justice*, Vol.4: 189–206.

Mesa, Darío. 1986. "La vida política después de Panamá", *Manual de Historia de Colombia*, T.III, 83–176. Bogotá: Instituto Colombiano de Cultura.

Minca, Claudio. 2007. "Agamben's geographies of modernity", *Political Geography*, Vol.26 (1): 78–97.

Minca, Claudio and Nick Vaughan-Williams. 2012. "Carl Schmitt and the Concept of the Border", *Geopolitics*, Vol.17 (4): 756–772.

Misión Colombia. 1988. *Historia de Bogotá, Siglo XX*, Vol. III. Bogotá: Villegas Editores.

Misiones Católicas del Putumayo: documentos relativos a esta Comisaría. 1914. Bogotá: Imprenta de San Bernardo.

Mitchell, Timothy. 2002a. *Rule of Experts. Egypt, techno-politics, modernity*. Berkeley, Los Angeles, London: University of California Press.

Mitchell, Timothy. 2002b. "The stage of modernity", in *Questions of Modernity*, edited by Timothy Mitchell, 1–34. Minneapolis & London: University of Minnesota Press.

Mitchell, Timothy. 2006. "Society, economy, and the state effect", in *The Anthropology of the State: A reader*, edited by Aradhana Sharma and Akhil Gupta, 169–186. Oxford: Blackwell.

Montclar, Fidel. 1924. *Conferencia leída por el Rvmo. P. Fr. Fidel de Montclar Prefecto Apostólico del Putumayo, en el Teatro Faenza, el día 20 de agosto de 1924, con motivo del Congreso Nacional de Misiones*. Bogotá: Casa Editorial Marconi.

Montenegro, Santiago. 2006. *Sociedad abierta, geografía y desarrollo. Ensayos de economía política*. Bogotá: Editorial Norma.

Montesquieu, Charles. 1990[1748]. "The spirit of the laws", in *Montesquieu, Selected Political Writings*, edited by Melvin Richter. Indianapolis, IN & Cambridge: Hackett.

Mora, Julio. 1997. *Mocoa. Su historia y desarrollo*. Bogotá: Imprenta Nacional.

Morales, Omar. 1997. *La Gesta de la Arriería*. Bogotá: Editorial Planeta.

Moran, Emilio. 1981. *Developing the Amazon: The social and ecological impact of settlement along the Transamazon highway*. Bloomington: Indiana University Press.

Múnera, Alfonso. 2005. *Fronteras imaginadas. La construcción de las razas y la geografía en el siglo XIX colombiano*. Bogotá: Editorial Planeta.

Nepstad, Daniel et al. 2001. "Road paving, fire regimes feedbacks, and the future of Amazon forests", *Forest Ecology and Management*, Vol.154 (3): 395–407.

Nieto, Mauricio. 2008. *Orden natural y orden social. Ciencia y política en el Semanario del Nuevo Reyno de Granada*. Bogotá: Uniandes-CESO.

Nishizaki, Yoshinori. 2008. "Suphanburi in the fast lane: Roads, prestige, and domination in provincial Thailand", *The Journal of Asian Studies*, Vol.67 (2): 433–467.

Nugent, David. 1994. "Building the state, making the nation: the bases and limits of state centralization in 'modern' Peru", *American Anthropologist*, Vol.96 (2): 333–369.

Oliveira, Adrilane and George H. de Moura. 2014. "Integração Regional, Desenvolvimento e Meio Ambiente: Impactos na Implementação de Projetos da IIRSA na Amazônia Brasileira", *Revista Eletrônica de Ciências Sociais, História e Relações Internacionais*, Vol.7 (1): 44–59.

Oliveira, Ariovaldo et al. 2005. *Amazonia Revelada. Os descaminos ao longo da BR-163*. Brasília: CNPq.

Ortiz, Sutti. 1984. "Colonization in the Colombian Amazon", in *Frontier Expansion in Amazonia*, edited by Marianne Schmink and Charles H. Hood, 204–230. Gainesville: University of Florida Press.

Pachón, Alvaro and María Teresa Ramírez. 2006. *La infraestructura de transporte en Colombia durante el siglo XX*. Bogotá: Fondo de Cultura Económica, Banco de la República.

Palacio, Germán. 2006. *Fiebre de tierra caliente. Una historia ambiental de Colombia 1850–1930*. Bogotá: Universidad Nacional de Colombia, ILSA.

Palacios, Marco. 1980. "La fragmentación regional de las clases dominantes en Colombia: Una perspectiva histórica", *Revista Mexicana de Sociología*, Vol.42 (4): 1663–1689.

Palacios, Marco. 2007. *Between Legitimacy and Violence. A history of Colombia, 1875–2002*. Durham, NC and London: Duke University Press.

Pandya, Vishvajit. 2002. "Contacts, images and imagination: The impact of a road in the Jarwa reserve forest, Andaman Islands", *Bijdragen tot de Taal-, Land-en Vokenkunde*, Vol.158 (4): 799–820.

Park, James William. 1985. *Rafael Núñez and the politics of Colombian regionalism, 1863–1886*. Baton Rouge & London: Louisiana State University Press.

Pécaut, Daniel. 1987. *Orden y Violencia: Colombia 1930–1954*. Bogotá: Cerec, Siglo XXI editores.

Pécaut, Daniel. 2000. "The loss of rights, the meaning of experience, and social connection: A consideration of the internally displaced in Colombia", *International Journal of Politics, Culture, and Society*, Vol.14 (1): 89–105.

Pécaut, Daniel. 2003. "El rostro ambiguo de Colombia", in *In-sur-gentes. Construir región desde abajo*, edited by William Torres, Luis Ernesto Lasso and Bernardo Tovar, 33–43. Neiva: Editorial Universidad Surcolombiana.

Perz, Stephen. 2014. "Sustainable development: The promise and perils of roads", *Nature*, Vol.513: 178–179.

Pina-Cabral, João. 1987. "Paved roads and enchanted Mooresses: The perception of the past among the peasant population of the Alto Minho", *Man*, Vol.22 (4): 715–735.

Pineda, Roberto. 1987. "El ciclo del caucho (1850–1932)", in *Colombia Amazónica*, 183–209. Bogotá: Universidad Nacional de Colombia, Fondo FEN.

Pineda, Roberto. 2003. "Vorágine o tierra de promisión. Trayectoria histórica de la Amazonia", in *In-sur-gentes. Construir región desde abajo*, edited by William Torres, Luis Ernesto Lasso and Bernardo Tovar, 143–178. Neiva: Editorial Universidad Surcolombiana.

Plácido, Fray. 1961. *Puerto Asís ayer y hoy: breves apuntes sobre su fundación y desarrollo, 1912–1962*. Sibundoy: Vicario Apostólico de Sibundoy.

Polo, José. 2010. "Las fronteras en la historia hispanoamericana: notas sobre algunas tendencias", *El Taller de la Historia*, Vol.2 (2): 206–224.

Pratt, Mary Louise. 1992. *Imperial Eyes. Travel writing and transculturation*. London: Routledge.

Presidencia de la República. 2005. *Discurso de Alvaro Uribe Vélez durante el Almuerzo ofrecido en honor al presidente de Brasil Luiz Inácio Lula da Silva*. Available from: http://www.presidencia.gov.co/prensa_new/sne/2005/diciembre/14/12142005.htm [Accessed 7th October 2016].

Quito, Jacinto M. 1938. *Miscelánea de mis treinta y cinco años de Misionero del Caquetá y Putumayo*. Bogotá: Editorial Aguila.

Ramírez, María Clemencia. 1996. *Frontera fluida entre Andes, Piedemonte y Selva: el caso del valle del Sibundoy, siglos XVI–XVIII*. Bogotá: Instituto Colombiano de Cultura Hispánica.

Ramírez, María Clemencia. 2001. *Entre el estado y la guerrilla: identidad y ciudadanía en el movimiento de los campesinos cocaleros del Putumayo*. Bogotá: ICANH, Colciencias.

Ramírez, María Clemencia. 2003. "El departamento del Putumayo en el contexto del suroccidente colombiano. Ordenamiento territorial y diferencias intrarregionales", in *In-sur-gentes. Construir región desde abajo*, edited by William Torres, Luis Ernesto Lasso and Bernardo Tovar, 203–239. Neiva: Editorial Universidad Surcolombiana.

Ramírez, María Clemencia. 2011. *Between the Guerrillas and the State: The Cocalero movement, citizenship, and identity in the Colombian Amazon*. Durham, NC: Duke University Press.

Ramírez, María Clemencia and Beatriz Alzate. 1995. "Por el Valle de Atriz a Ecija de Sucumbíos", in *Caminos Reales de Colombia*, edited by Mariano Useche. Bogotá: Fondo FEN.

Ramírez, María Clemencia et al. 2010. *Elecciones, coca, conflicto y partidos políticos en Putumayo 1980–2007*. Bogotá: Icanh, Cinep.

Rappaport, Joanne. 1998. *The Politics of Memory. Native historical interpretation in the Colombian Andes*. Durham, NC: Duke University Press.

Ratzel, Friedrich. 1896. "The territorial growth of states", *Scottish Geographical Magazine*, Vol.12 (7): 351–361.

Rausch, Jane. 1999. *La Frontera de los Llanos en la historia de Colombia (1830–1930)*. Bogotá: Banco de la República, El Áncora Editores.

Rausch, Jane. 2003. "La mirada desde la periferia: desarrollos en la historia de la frontera colombiana desde 1970 hasta el presente", *Fronteras de la Historia*, Vol.8: 251–260.

Reichel-Dolmatoff, Gerardo. 1965. *Colombia. Ancient peoples and places*. London: Thames and Hudson.

Revelo, Guido. 2006. *Puerto Asís. Una aproximación a su historia entre los años 1912 y 1960*. Bucaramanga: Editorial SIC.

Revista Semana. 2008. Falsos positivos mortales. 27th September. Available from: http://www.semana.com/nacion/articulo/falsos-positivos-mortales/95607-3 [Accessed 7th October 2016].

Reyes, Rafael. 1902. *A través de la América del Sur. Exploraciones de los hermanos Reyes*. México: Ramón de S.N. Araluce.

Reyes, Rafael. 1912. *Por Colombia, por Ibero-América*. London: Imprenta de Wertheimer, LEA & Cia.

Reyes, Rafael. 1913. Carta del General Reyes. *El Nuevo Tiempo*, Bogotá, 21st November 1913.

Reyes, Rafael. 1914. *The Two Americas*. New York: Frederick A. Stokes Company.

Reyes, Rafael. 1920. *Escritos varios*. Bogotá: Tipografía Ancovar.

Reyes, Rafael. 1979 [1902]. *Across the South American continent. Explorations of the brothers Reyes*. Bogotá: Flota Mercante Grancolombiana.

Reyes, Rafael. 1986. *Memorias 1850–1885*. Bogotá: Fondo Cultural Cafetero.

Rippy, J. Fred. 1943. "Dawn of the railway era in Colombia", *The Hispanic American Historical Review*, Vol.23 (4): 650–663.

Romo, Franco. 1990. *Carreteras Variantes*. Pasto: no publisher.

Roseberry, William. 2004. "Hegemony and the language of contention", in *Everyday Forms of State Formation. Revolution and the negotiation of rule in Mexico*, edited by Gilbert Joseph and Daniel Nugent, 355–366. Durham, NC: Duke University Press.

Roseman, Sharon. 1996. "'How we built the road': The politics of memory in rural Galicia", *American Ethnologist*, Vol.23 (4): 836–860.

Rousseau, Jean-Jacques. 2004[1762]. *The Social Contract*. London: Penguin.

Safford, Frank. 1991. "Race, integration, and progress: Elite attitudes and the Indian in Colombia, 1750–1870", *Hispanic American Historical Review*, Vol.71 (1): 1–33.

Safford, Frank. 2010. "El problema de los transportes en Colombia en el siglo XIX", in *Economía Colombiana del Siglo XIX*, edited by Adolfo Meisel and María Teresa Ramírez, 523–573. Bogotá: Banco de la República and Fondo de Cultura Económica.

Safford, Frank and Marco Palacios. 2002. *Colombia: Fragmented land, divided society*. Oxford: Oxford University Press.

Salamanca, Demetrio. 1994. *La Amazonia Colombiana*, vol.2. Tunja: Academia Boyacense de Historia.

Salazar, Carolina and Vince McElhinny. 2008. "Carretera Pasto Mocoa – Colombia: BID Compromiso a la Sostenibilidad en la Mira". BICECA. Available from http://www.bicusa.org/es/Article.10935.aspx [Accessed: 10th January 2009].

Salazar, Jaime. 2000. *De la mula al camión. Apuntes para una historia del transporte en Colombia*. Bogotá: TM Editores.

Sánchez, Efraín. 1998. *Gobierno y geografía. Agustín Codazzi y la Comisión Corográfica de la Nueva Granada*. Bogotá: Banco de la República, El Ancora Editores.

Sánchez, Lina María. 2007. *Impacto urbano del desplazamiento forzado en Mocoa-Putumayo. Elementos de diagnóstico y planteamientos para un reordenamiento espacial*. Bogotá: CINEP.

Sánchez, Ricardo, ed. 1908. *La Reconstrucción Nacional. Estudio de la administración del excelentísimo General D. Rafael Reyes*. Bogotá: Casa Editorial La Prensa.

Sanford, Victoria. 2004. "Contesting displacement in Colombia. Citizenship and state sovereignty at the margins", in *Anthropology in the Margins of the State*, edited by Veena Das and Deborah Poole, 253–277. Santa Fe: School of American Research Press.

Schmink, Marianne. 1982. "Land conflicts in Amazonia", *American Ethnologist*, Vol.9(2): 341–357.

Schmink, Marianne and Charles H. Hood, eds. 1984. *Frontier Expansion in Amazonia*. Gainesville: University of Florida Press.

Schmitt, Carl. 1985[1922]. *Political Theology: Four chapters on the concept of sovereignty*. Cambridge, MA: MIT Press.

Schmitt, Carl. 2006[1950]. *The Nomos of the Earth in the International Law of the Jus Publicum Europaeum*, translated and annotated by G. L. Ulmen. New York: Telos Press Publishing.

Scott, James. 1998. *Seeing Like a State. How certain schemes to improve the human condition have failed*. New Haven, CT & London: Yale University Press.

Scott, James. 2009. *The Art of Not Being Governed. An anarchist history of upland Southeast Asia*. New Haven, CT & London: Yale University Press.

Segura, Nora. 2002. "Colombia: A New Century, an Old War, and More Internal Displacement", *International Journal of Politics, Culture, and Society*, Vol.14(1): 107–126.

Selwyn, Tom. 2001. "Landscapes of separation: reflections in the symbolism of by-pass roads in Palestine", in *Contested Landscapes. Movement, exile and place*, edited by Barbara Bender and Margot Winer, 225–240. Oxford & New York: Berg.

Serje, Margarita. 2011. *El revés de la nación. Territorios salvajes, fronteras y tierras de nadie*. Bogotá: Universidad de los Andes-CESO.

Sharma, Aradhana and Akhil Gupta, eds. 2006. *The Anthropology of the State: A reader*. Oxford: Blackwell.

Shaw, Ian and Majed Akhter. 2012. "The unbearable humanness of drone warfare in FATA, Pakistan", *Antipode*, Vol.44 (4): 1490–1509.

Shenhav, Yehouda. 2012. "Imperialism, exceptionalism and the contemporary world", in *Agamben and Colonialism*, edited by Marcelo Svirsky and Simone Bignall, 17–31. Edinburgh: Edinburgh University Press.

Stanfield, Michael Edward. 1998. *Red Rubber, Bleeding Trees: Violence, slavery, and empire in northwest Amazonia, 1850–1933*. Albuquerque: University of New Mexico Press.

Stepan, Nancy. 1991. *The Hour of Eugenics: Race, gender, and nation in Latin America*. Ithaca, NY: Cornell University Press.

Stewart, Douglas I. 1994. *After the Trees: Living on the Transamazon highway*. Austin: University of Texas Press.

Sundberg, Juanita. 2015. "The state of exception and the imperial way of life in the United States–Mexico borderlands", *Environment and Planning D: Society and Space*, Vol.33: 209–228.

Svirsky, Marcelo and Simone Bignall, eds. 2012. *Agamben and Colonialism*. Edinburgh: Edinburgh University Press.

Taussig, Michael. 1991. *Shamanism, Colonialism, and the Wild Man. A study on terror and healing*. Chicago: The University of Chicago Press.

Taussig, Michael. 1992. *The Nervous System*. London & New York: Routledge.

Taussig, Michael. 1997. *The Magic of the State*. London & New York: Routledge.

The New York Times. 1914. "Two Americas. Travels north and south by Gen. Reyes." 8th March. Available from: http://query.nytimes.com/mem/archive-free/pdf?res=F10E12FC3D5E13738DDDA10894DB405B848DF1D3 [Accessed 7th October 2016].

Thevenot, Laurent. 2002. "Which road to follow? The moral complexity of an 'equipped' humanity", in *Complexities: Social studies of knowledge practices*, edited by John Law and Annemarie Mol, 53–87. Durham, NC and London: Duke University Press.

Thomas, Philip. 2002. "The river, the road, and the rural-urban divide: A post-colonial moral geography from southeast Madagascar", *American Ethnologist*, Vol.29, 2: 366–391.

Torres, María Clara. 2007. "Comunidades y coca en el Putumayo: prácticas que hacen aparecer al estado", *Controversia, nro. 188*. Bogotá: CINEP.

Torres, María Clara. 2011. *Estado y coca en la frontera colombiana. El caso del Putumayo*. Bogotá: Odecofi-CINEP.

Townsend, Janet. 1977. "Perceived worlds of the colonist of tropical rainforest, Colombia", *Transactions of the Institute of British Geographers*, New Series Vol.2 (4): 430–458.

Triana, Miguel. 1950[1907]. *Por el sur de Colombia. Excursión pintoresca y científica al Putumayo*. Bogotá: Prensa del Ministerio de Educación Nacional.

Trouillot, Michel-Rolph. 1995. *Silencing the past. Power and the production of history*. Boston, MA: Beacon Press.

Trouillot, Michel-Rolph. 2001. "The anthropology of the state in the age of globalization. Close encounters of a deceptive kind", *Current Anthropology*, Vol.42 (1): 125–138.

Turner, Frederick J. 2008[1893]. "The significance of the frontier in American History", in *The Frontier in American History*, Frederick J. Turner, 13–42. Charleston, SC: Bibliobazaar.

UNHRC. 2010. "Report of the Special Rapporteur on extrajudicial, summary or arbitrary executions, Philip Alston". Available from: http://www2.ohchr.org/english/bodies/hrcouncil/docs/14session/A.HRC.14.24.Add.2_en.pdf [Accessed 7th October 2016].

UNHCR. 2015. Global trends. Forced displacement in 2015. Available from: http://www.unhcr.org/statistics/unhcrstats/576408cd7/unhcr-global-trends-2015.html [Accessed 10th October 2016].

UNODC. 2006. "Cultivos de Coca. Estadísticas municipales. Censo 2006". Available from: ftp://190.144.33.2/UNODC/municipios_2006.pdf [Accessed 10th December 2011].

UNODC. 2008. *Colombia. Monitoreo de Cultivos de Coca*. Bogotá: UNODC, Gobierno de Colombia.

UNODC and Acción Social. 2007. Informe ejecutivo sobre el seguimiento a los programas de Familias Guardabosques y Proyectos Productivos. Available from: ftp://190.144.33.2/UNODC/informeejecutivo.pdf [Accessed: 10th December, 2011].

Uribe, María Teresa. 2001. *Nación, ciudadano y soberano*. Medellín: Corporación Región.

Uribe, María Victoria. 1986. "Pastos y protopastos: la red regional de intercambio de productos y materias primas de los siglos X a XVI D.C.", *Revista Maguaré*, Vol.3: 33–43.

Uribe, María Victoria. 1995. "Caminos de los Andes del Sur. Los caminos del sur del Cauca y Nariño", in *Caminos Reales de Colombia*, edited by Mariano Useche. Bogotá: Fondo FEN.

Uribe, Simón. 2015. "Construyendo el trópico: relatos de viajeros ingleses en Colombia durante el siglo XIX", in *Semillas de Historia Ambiental*, edited by Stefania Gallini, 215–249. Bogotá: Jardín Botánico José Celestino Mutis, Universidad Nacional de Colombia.

Vásquez, María de la Luz. 2006. "De repúblicas independientes a zona de despeje. Identidades y estado en los márgenes", in *Identidades culturales y formación del Estado en Colombia. Colonización, naturaleza y cultura*, edited by Ingrid Bolívar, 119–207. Bogotá: Universidad de los Andes, CESO.

Vélez, Humberto. 1989. "Rafael Reyes: Quinquenio, régimen político y capitalismo (1904–1909)", in *Nueva Historia de Colombia*, edited by Alvaro Tirado, Vol. I, 187–214. Bogotá: Editorial Planeta.

Vilanova, Pacífico. 1947. *Capuchinos Catalanes en el sur de Colombia*, 2 vols. Barcelona: Imprenta Myria.

Villava, Angel. 1895. *Una visita al Caquetá por un misionero capuchino*. Barcelona: Librería y tipografía Católica.

Wade, Peter. 1989. "The language of race, place and nation in Colombia", *América Negra*, 2: 41–68.

Wade, Peter. 1993. *Blackness and Race Mixture: The dynamics of racial identity in Colombia*. Baltimore, MD: Johns Hopkins University Press.

Wade, Peter. 1997. *Race and Ethnicity in Latin America*. London and Chicago: Pluto Press.

Weber, David. 1986. "Turner, the Boltonians, and the borderlands", *The American Historical Review*, Vol.91 (1): 66–81.

Weber, David and Jane Rausch, eds. 1994. *Where Cultures Meet. Frontiers in Latin American history*. Wilmington, DE: Jaguar Books.

Weber, Max. 1998. "Politics as vocation", in *From Max* Weber, edited by H.H. Gerth and C.W. Mills. London: Routledge.

Wesche, Rolf. 1974. *El Desarrollo del poblamiento en el alto valle del río Putumayo*. Bogotá: IGAG.

Wilson, Fiona. 2004. "Towards a political economy of roads: Experiences from Peru", *Development and Change*, Vol.35 (3): 525–546.

Wylie, Lesley. 2013. *Colombia's Forgotten Frontier: A literary geography of the Putumayo*. Liverpool: Liverpool University Press.

Zárate, Carlos. 2001. *Extracción de quina: la configuración del espacio Andino-Amazónico de fines de siglo XIX*. Bogotá: Universidad Nacional de Colombia.

Zárate, Carlos. 2008. *Silvícolas, siringueros y agentes estatales. El surgimiento de una sociedad transfronteriza en la Amazonia de Brasil, Perú y Colombia 1880–1932*. Bogotá: Universidad Nacional de Colombia, Imani.

Index

Frontier Road: Power, History, and the Everyday State in the Colombian Amazon,
First Edition. Simón Uribe.
© 2017 John Wiley & Sons Ltd. Published 2017 by John Wiley & Sons Ltd.